OPTICAL BISTABILITY, INSTABILITY AND OPTICAL COMPUTING

Proceedings of the 1987 International Topic Meeting on
'Optical Bistability, Instability and Optical Computing'

OPTICAL BISTABILITY, INSTABILITY AND OPTICAL COMPUTING

Beijing, People's Republic of China
24-29 Aug 1987

Editors
H-Y ZHANG
KK LEE

Published by

World Scientific Publishing Co. Pte. Ltd.
P.O. Box 128, Farrer Road, Singapore 9128

U. S. A. office: World Scientific Publishing Co., Inc.
687 Hartwell Street, Teaneck NJ 07666, USA

Library of Congress Cataloging-in-Publication data is available.

OPTICAL BISTABILITY, INSTABILITY AND OPTICAL COMPUTING

ISBN 9971-50-547-9
9971-50-578-9 (pbk)

Printed in Singapore by Utopia Press.

PREFACE

The International Topic Meeting on Optical Bistability Instability and Optical Computing hosted by Beijing Institute of Modern Physics, Peking University, Beijing, P.R.China from August 24-29, 1987. About 100 participates attended the Topic Meeting. There are 8 invited talks and 5 more Chinese scientists were invited to give talks about the research works on O.B. and optical computing of their institutes. 41 papers were collected in this proceeding. The panel discussion was held during the meeting. The summary of the panel discussion was also collected.

We are grateful to the invited speakers, presenters of the contributed papers and other participants for their contribution and participation. We feel that this meeting is quite successful in attaining its goal of providing mutual exchange among colleagues as well as providing personal interactions. We also hope that this proceeding will be of some use for those scientists and students who are interested in working on these interdisciplinary topics.

This Topic Meeting is the first in these topics in China. We sincerely hope that this will be the first of a series. The themes of the future meeting in this series will be more focused.

We are grateful to our sponsors: International Centre of Theoretical Physics (ICTP) and Natural Science Foundation of China for their generous finacial support. We would like to express our appreciation to Professor He Xue-hua, Drs. Hou Fu-xing, Wang Wei-li and Chen Wei-xi for their administrative arrangements, and Chen Hong-pei, Liu Yue for their taking part in the edi[illegible]orks. Last but not the least, we would like to ac[illegible] the assistance and advice given to us by the Publisher, in particular, the Editor-in-Chief, Dr. Pan Guo-ju.

Zhang He-yi

Kotik K. Lee

Leadership of the Meeting

INTERNATIONAL ADVISORY COMMITTEE

Prof. Wang Da-heng	Optical Society of China
Prof. H. M. Gibbs	Optical Science Center, University of Arizona
Prof. Wang Zhi-jiang	Shanghai Institute of Optical and Fine Mechanics

ORGANIZING COMMITTEE

Chairman	Prof. Zhang He-yi Beijing Institute of Modern Physics	
Co-Chairman	Dr. Kotik K. Lee Perkin-Elmer Co., U.S.A.	
Members		
	Prof. Zhang Hong-jun	Physics Institute Academia Sinica
	Prof. Li Chun-fei	Harbin Institute of Technology
	Prof. Zhang Ke-qian	Tsing Hua University
	Prof. Wang Qi-ming	Institute of Semiconductor Academia Sinica
	Associate Prof. Xu Ke-shou	Fudan University
Secretary	Dr. Hou Fu-xing	Dept. of Phys., Peking Univ.
	Dr. Wang Wei-li	Dept. of Phys., Peking Univ.

INTERNATIONAL TOPIC MEETING ON OPTICAL BISTABILITY INSTABILITY AND OPTICAL COMPUTING

BEIJING CHINA 24-29 AUGUST 1987

ORGANIZED BY
BEIJING INSTITUTE OF MORDERN PHYSICS
PEKING UNIVERSITY

SPONSORED BY
BEIJING INSTITUTE OF MORDERN PHYSICS
PHYSICS INSTITUTE ACADEMIA SINICA
PHYSICS DEPARTMENT OF HARBIN INSTITUTE OF TECHNOLOGY
DEPT. OF INFORMATION ELECTRONICS TSING HUA UNIV.
INSTITUTE OF SEMICONDUCTOR ACADEMIA SINICA
DEPARTMENT OF PHYSICS FUDAN UNIVESITY

SUPPORTED BY
ICTP
NATURAL SCIENCE FOUNDATION OF CHINA
ALL SPONSORED INSTITUTES

CONTENT

OPTICAL BISTABILITY AND NONLINEAR OPTICS (2)

THEORY

OPTICAL INSTABILITY AND CHAOS

OPTICAL COMPUTING

MISCELLANY

Optical Bistability in GaAs Etalons and Waveguides

H. M. Gibbs, G. Khitrova, S. Koch, N. Peyghambarian, D. Sarid, A. Chavez-Pirson, W. Gibbons, A. Jeffery, K. Komatsu,† Y. H. Lee, D. Hendricks,†† J. Morhange,* S. H. Park, M. Warren

A. C. Gossard and W. Wiegmann, AT&T Bell Labs, Murray Hill, NJ 07974

M. Sugimoto, Opto-Electronics Research Laboratories, NEC

Abstract

Nonlinear absorption spectra of bulk GaAs and 299-, 152-, and 76-Å GaAs/AlGaAs multiple-quantum-well (MQW) samples are reported. Nonlinear index spectra are calculated by Kramers-Kronig transformations of the absorption spectra. It is concluded that the principal nonlinear refractive mechanisms for optical bistability in bulk GaAs etalons are band-filling and reduction of the Coulomb enhancement of continuum states; exciton saturation and band-gap renormalization occur at lower carrier densities, but their contributions are of opposite sign and almost cancel. Exciton saturation nonlinearities dominate for bistability in MQW samples with <100-Å wells.

Optical bistability is reported in a strip-loaded waveguide formed by reactive ion etching; the guiding layer consisted of 60 periods of 100-Å GaAs and 100-Å $Al_{0.28}Ga_{0.72}As$. The power for bistability is about the same in the etalon and in the waveguide. The 100-fold increase in length of the nonlinear material from the etalon to the waveguide requires a greater detuning to keep the absorption acceptably low, resulting in lower carrier density and a weaker nonlinear refractive effect per unit length. The cleaved ends of the waveguide were uncoated.

Present addresses:
†NEC Corporation, 1-1 Miyazaki 4-Chome, Miyamae-Ku Kawasaki, Kanagawa 213, Japan
††1058 E. 17th Street, Salt Lake City, UT 84105
*Laboratoire de Physique des Solides, Universite P. et M. Curie, 4 Place Jussieu, 75230 Paris Cedex 05, France

This article is restricted to a discussion of optical nonlinearities and logic operations using GaAs materials. For completeness we give reference to our other work related to this conference. ZnS interference filters, first demonstrated to be optically bistable by Karpushko and Sinitsyn,[1] have been used as AND and OR gates to demonstrate one-bit addition by symbolic substitution[2] and to recognize three-spot patterns in the shape of a V or Γ in a 2×9 input array.[3] These simple demonstrations of nonlinear decision-making[4] devices to parallel optical computing illustrate logic operations, cascading, and a small amount of parallelism. Observations of optical instabilities in sodium vapor using a single feedback mirror[5] arise from transverse effects and must be related to the hybrid-system transverse instabilities reported at this conference by Worontsov and Shmalhausen.

In GaAs we have shown that the AC Stark shift of the exciton can be used to shift a Fabry-Perot etalon peak just below the exciton to perform an optical NOR gate; its response is determined by the pump pulse's duration.[6] However, this gate does not have gain; i.e., the energy of the pulse being controlled is less than the pump-pulse energy, unlike the NOR gate based on above-bandgap absorption of the input pulses.[7] But the response time of the latter gate is determined by the lifetime of the carriers, typically several nanoseconds[7] but as low as ≅50 ps in 0.3-μm-thick GaAs with no AlGaAs windows.[8] This two-wavelength operation complicates cascading because the output is at a different wavelength from the inputs. One-wavelength bistable or transistor operation[4] avoids this complication, but gain with picosecond pulses has not been shown experimentally. Here we focus attention on bistable operation just below the band edge using 100 ns to 1 μs pulses, short enough to avoid thermal effects and long enough that the carriers are in thermodynamic equilibrium under the action of the pump light and relaxation processes.

The band-edge optical nonlinearity measurements[9] were motivated by the great similarity[10] between optical bistability (switch-up powers and hysteresis loops) using bulk or MQW etalons. The photoluminescence from a 299-Å MQW platelet excited by 514.5 nm 0.8-μs light served as a broad-band probe source, so that absorption data could be taken simultaneously over the band edge; see Fig. 1a. The 821-nm, 1-μs pump pulses were focused to about a 15-μm-diameter spot, larger than the probe. These data are compared with the Banyai-Koch[11] plasma theory which is based on the Haug and Schmidt-Rink[12] many body theory of band-edge nonlinearities of direct-gap semiconductors. Included in the theory are band filling and plasma screening of the Coulomb interaction leading to three other mechanisms, namely bandgap renormalization, exciton saturation, and reduction of the enhancement of the continuum states (which, of course, makes the band edge square rather than parabolic). Because of the long duration of the pulses and the resultant quasi-equilibrium in the system of electronic excitations, one can calculate the corresponding, intensity dependence of the index of refraction by making a Kramers-Kronig calculation:

$$\Delta n = \frac{\hbar c}{\pi} P \int_{E_1}^{E_2} \frac{\Delta\alpha(E')}{E'^2 - E^2} dE' .$$

See Fig. 1c and d; Fig. 1c is consistent with direct measurements.[13] The clear similarity between the data and the theory, give us the confidence to identify the principal mechanisms leading to bistability; in fact one can see a clear discrepancy between the data and theory if any one of exciton bound states or band filling or bandgap renormalization is neglected. At low intensities exciton saturation and bandgap renormalization dominate, but their index contributions have opposite signs and very nearly cancel. The multiple-quantum-well data are handled similarly, but the theory is not so well developed for the realistic case of quantum wells with a finite thickness, which are neither three dimensional or two dimensional. The imperical technique of Chemla and Miller[14] was used in which the band-edge absorption is fitted to the sum of Gaussians for the heavy- and light-hole excitonic transitions and for the band-to-band transitions; see Fig. 2. The maximum index change at each intensity is shown in Fig. 3 for bulk GaAs and for MQW with 299-Å, 152-Å, and 76-Å wells. The index change is seen to increase with narrower wells. However, the pump absorption was different for the various samples; Figure 4 shows the maximum change in index per carrier, $\Delta n/N$, versus the carrier concentration N. Clearly this ratio is almost the same in all four samples as long as one assumes that the carrier lifetime is the same (20 ns) in all four samples. But even if the carrier lifetimes turn out to be unequal, these results point to why bistability looks so similar in bulk and MQW etalons. The best operating point must be close to the wavelength at which the linear absorption αL equals the mirror transmissivity $T=1-R$. Then if the absorption is the same in two samples, so will be the carrier density and consequently the index changes and bistability. However, we stress that the mechanisms are different (at room temperature): excitonic in MQW and a combination of band filling and reduction of Coulomb enhancement in bulk GaAs.

Clearly much of the motivation for understanding and optimizing GaAs nonlinearities and bistability is the hope that someday arrays of such devices will play a major role in parallel computing, associative memories, learning machines, etc. Reactive ion etching of arrays is relatively easy.[15] For example, 10^4 NOR gates on 4 cm^2 operated at 1 GHz and requiring 40 pJ per bit operation, would require 400 W of average laser power and produce 100 W/cm^2 of heat. But the system would perform 10^{13} bit operations per second -- more than a CRAY.

In many applications in optical communications, interconnects, multiplexing, encryption, etc., the data are handled sequentially in a pipeline fashion. In such cases waveguide devices could be used easily, assuming performance superior to etalons. In quest of such, we have used reactive ion etching to fabricate strip-loaded waveguides in which electronic optical bistability has been seen.[16] Of course, this is not the first observation of bistability in a waveguide. Thermal bistability has been seen in prism coupling into ZnS and ZnSe slabs,[17,18] increasing absorption and dispersive in Fabry-Perot-like GaAs MQW slab,[19] and increasing absorption in a ZnSe slab.[20] Hybrid self-electrooptic-effect-device bistability was seen in a GaAs MQW slab.[21] By far the closest to our work is that in a dispersive Fabry-Perot-like GaAs MQW channel guide in which a Si_3N_4 strip on top strains the MQW at its edges resulting in guiding.[22] Our guide has yielded better bistability (wider loops and sharper switching) and should be easier to model and to fabricate reproducibly. The sample was grown at NEC using a GaAs substrate, 3-μm-thick $Al_{0.3}Ga_{0.7}As$ layer to reduce losses into the substrate, then 60 periods of 100-Å GaAs and 100-Å $Al_{0.28}Ga_{0.72}As$, and finally 1-μm-thick $Al_{0.13}Ga_{0.87}As$ on top. A 3-μm-wide strip was etched about 0.9 μm into the top layer, resulting in lateral guiding in the MQW layer so that wall roughness in the top strip is less significant. A typical guide is shown in Fig. 5. Optical bistability in a 200-μm-long guide with 300-ns input pulses is shown in Fig. 6. Thermal effects have been ruled out by checking that the laser frequency is above the Fabry-Perot peak. With longer pulses ($\cong$8 μs) heating in the upper branch results in switch-down at a higher power than switch-up. Multiple bistability consisting of two bistability loops shows that a 2π phase shift has been achieved in this waveguide compared with about 0.5π in etalons.

However, in some respects the waveguide bistability is disappointing. The usual argument is that a waveguide keeps the light at a high intensity over a longer interaction length so that the input power can be reduced. This has shown to be the case for many nonlinear effects in transparent materials in which the absorption length α^{-1} greatly exceeds the Rayleigh length of a tightly focused beam in free space. But that is not the situation in GaAs etalons where one operates as close to the band edge as the finesse will allow ($\alpha L \cong T$). Increasing the interaction length by 100 or more must be accompanied by increased detuning from the band edge to preserve $\alpha L \cong T$. The similarity in operating powers between MQW GaAs bistability in etalons and waveguides suggests that the nonlinear refraction decreases to compensate for the increased interaction length. On a more optimistic note, this first attempt used only cleaved end faces ($R \cong 0.3$) and has not been optimized. Ultimately, one hopes for picosecond[6] response times in a variety of waveguide devices such as directional couplers Mach-Zehnder interferometers, etc. The devices reported here are a first step in a material compatible with integrated electronics and optoelectronics.

In summary, bulk GaAs and MQW etalons exhibit similar optical bistability powers and loops, but band filling and reduction of the Coulomb enhancement of continuum states are the nonlinear refraction mechanisms in bulk GaAs and exciton saturation effects in MQW. The tunability of the band edge with well thickness in a MQW is very useful for diode laser operation. Bistability has been seen in a MQW strip-loaded waveguide, but the power required is similar to that for etalon bistability.

We gratefully acknowledge support from the AFOSR, ARO, NSF, DARPA/RADC, SDI, and the Optical Circuitry Cooperative.

References

1. F. V. Karpushko and G. V. Sinitsyn, Appl. Phys. B **28**, 137 (1982) and J. Appl. Spectrosc. USSR **29**, 1323 (1978).

2. M. T. Tsao, L. Wang, R. Jin, R. W. Sprague, G. Gigioli, H.-M Kulcke, Y. D. Li, H. M. Chou, H. M. Gibbs, and N. Peyghambarian, Opt. Eng. **26**, 41(1987). L. Wang, H. M. Chou, H. M. Gibbs, G. C. Gigioli, G. Khitrova, H.-M. Kulcke, R. Jin, H. A. Macleod, N. Peyghambarian, R. W. Sprague, and M. T. Tsao, SPIE O-E LASE '87.

3. H. M. Gibbs, G. Khitrova, L. Wang, et al., OSA Second Topical Meeting on Optical Computing, Lake Tahoe, March 1987.

4. H. M. Gibbs and N. Peyghambarian in SPIE **634**, *Optical and Hybrid Computing*, 142 (1986). H. M. Gibbs, *Optical Bistability: Controlling Light with Light* (Academic, New York, 1985).

5. H. M. Gibbs, M. W. Derstine, K. Tai, J. F. Valley, J. V. Moloney, F. A. Hopf, M. Le Berre, E. Ressayre, and A. Tallet, in *Optical Instabilities*, R. W. Boyd, M. G. Raymer, and L. M. Narducci, eds. (Cambridge University, Cambridge, 1986) p. 340. G. Giusfredi, J. F. Valley, R. Pon, G. Khitrova, H. M. Gibbs, Int'l. Workshop on Instabilities, Dynamics, and Chaos in Nonlinear Optical Systems, Lucca, Italy, July 8-10, 1987.

6. D. Hulin, A. Mysyrowicz, A. Antonetti, A. Migus, W. T. Masselink, H. Morkoc, H. M. Gibbs, and N. Peyghambarian, Appl. Phys. Lett. **49**, 749 (1986). A. Mysyrowicz, D. Hulin, A. Migus, A. Antonetti, H. M. Gibbs, N. Peyghambarian, and H. Morkoc, OSA Topical Meeting on Picosecond Electronics and Optoelectronics, Lake Tahoe, January 1987.

7. J. L. Jewell, Y. H. Lee, M. Warren, H. M. Gibbs, N. Peyghambarian, A. C. Gossard, and W. Wiegmann, Appl. Phys. Lett. **46**, 918 (1985).

8. Y. H. Lee, H. M. Gibbs, J. L. Jewell, J. F. Duffy, T. Venkatesan, A. C. Gossard, W. Wiegmann, and J. H. English, Appl. Phys. Lett. **49**, 486 (1986).

9. Y. H. Lee, A. Chavez-Pirson, S. W. Koch, H. M. Gibbs, S. H. Park, J. Morhange, A. Jeffery, N. Peyghambarian, L. Banyai, A. C. Gossard, and W. Wiegmann, Phys. Rev. Lett. **57**, 2446 (1986).

10. S. Ovadia, H. M. Gibbs, J. L. Jewell, and N. Peyghambarian, Opt. Eng. **24**, 565 (1985).

11. L. Banyai and S. Koch, Z. Phys. B **63**, 283 (1986).

12. H. Haug and S. Schmitt-Rink, Prog. Quantum Electron. **9**, 3 (1984). H. Haug in: "Nonlinear Optical Properties of Semiconductors," ed. H. Haug (Academic, New York, 1987).

13. Y. H. Lee, A. Chavez-Pirson, B. K. Rhee, H. M. Gibbs, A. C. Gossard, and W. Wiegmann, Appl. Phys. Lett. **49**, 1505 (1986).

14. D. S. Chemla and D. A. B. Miller, J. Opt. Soc. Am. B 2, 1155 (1985). D. S. Chemla, D. A. B. Miller, P. W. Smith, A. C. Gossard, and W. Wiegmann, IEEE J. Quantum Electron. **QE-20**, 265 (1984).

15. T. Venkatesan, B. Wilkens, M. Warren, Y. H. Lee, G. Olbright, H. M. Gibbs, N. Peyghambarian, J. S. Smith, and A. Yariv, Appl. Phys. Lett. 48, 145 (1986). Y. H. Lee, M. Warren, G. R. Olbright, H. M. Gibbs, N. Peyghambarian, T. Venkatesan, J. S. Smith, and A. Yariv, Appl. Phys. Lett. 48, 754 (1986).

16. M. Warren, W. Gibbons, K. Komatsu, D. Sarid, D. Hendricks, H. Gibbs, and M. Sugimoto, IQEC Baltimore (1987), Postdeadline PD9.

17. W. Lukosz, P. Pirani, and V. Briguet, in *Optical Bistability III*, H. M. Gibbs, P. Mandel, N. Peyghambarian, and S. D. Smith, eds. (Springer-Verlag, Berlin, 1986) p. 108.

18. G. Assanto, B. Svensson, D. Kuchibhatla, U. J. Gibson, C. T. Seaton, and G. I. Stegeman, Opt. Lett. 11, 644 (1986).

19. A. C. Walker, J. S. Aitchinson, S. Ritchie, and P. M. Rodgers, Electron. Lett. 22, 366 (1986).

20. B. Y. Kim, Elsa M. Garmire, N. Shibata, and S. Zembutsu, CLEO, Baltimore, 1987.

21. J. S. Weiner, D. A. B. Miller, D. S. Chemla, T. C. Damen, C. A. Burrus, T. H. Wood, A. C. Gossard, and W. Weigmann, Appl. Phys. Lett. 47, 1148 (1985).

22. P. Li Kam Wa, P. N. Robson, J. P. R. David, G. Hill, P. Mistry, M. A. Pate, and J. S. Roberts, Electron. Lett. 22, 1129 (1986).

Figure Captions

Fig. 1. Room-temperature bulk GaAs optical nonlinearities: experiment and theory. (a) Experimental absorption spectra for different excitation intensities I (mW): 1) 0; 2) 0.2; 3) 0.5; 4) 1.3; 5) 3.2; 6) 8; 7) 20; 8) 50 on a 15-μm-diameter spot. (b) Nonlinear refractive index changes corresponding to the measured absorption spectra. The curves (a-g) in Fig. 1(b) are obtained by the Kramers-Kronig transformation of the corresponding experimental data (2-8) in Fig. 1(a). (c) Calculated absorption spectra for different electron-hole pair densities N (cm^{-3}): 1) 10^{15}; 2) 8×10^{16}; 3) 2×10^{17}; 4) 5×10^{17}; 5) 8×10^{17}; 6) 10^{18}; 7) 1.5×10^{18}. E_g^0 = 1.433 eV and E_R = 4.2 meV. (d) Calculated nonlinear refractive index changes. The curves (a-f) in Fig. 1(d) are obtained from the curves (2-7) in Fig. 1(c), respectively.

Fig. 2. 76-Å MQW linear absorption and fit to two Gaussians, simulating the heavy-hole and light-hole excitons, together with a broadened two-dimensional continuum (Sommerfeld factor), to simulate the band-to-band transitions.

Fig. 3. Maximum change in refractive index Δn as a function of input light intensity I.

Fig. 4. Maximum change in index per carrier versus the electron-hole concentration.

Fig. 5. Electron micrograph of GaAs/AlGaAs strip-loaded waveguide.

Fig. 6. Experimental input-output behavior of a bistable MQW waveguide operating at 867 nm. The upper trace is the output pulse and the lower trace is the input. The trace in the lower left is a plot of output versus input. Input power: 130 mW; output power: 15 mW; estimated power coupled into waveguide: 30 mW; band-edge wavelength: 839 nm; αL = 0.575, including both guiding and absorptive losses.

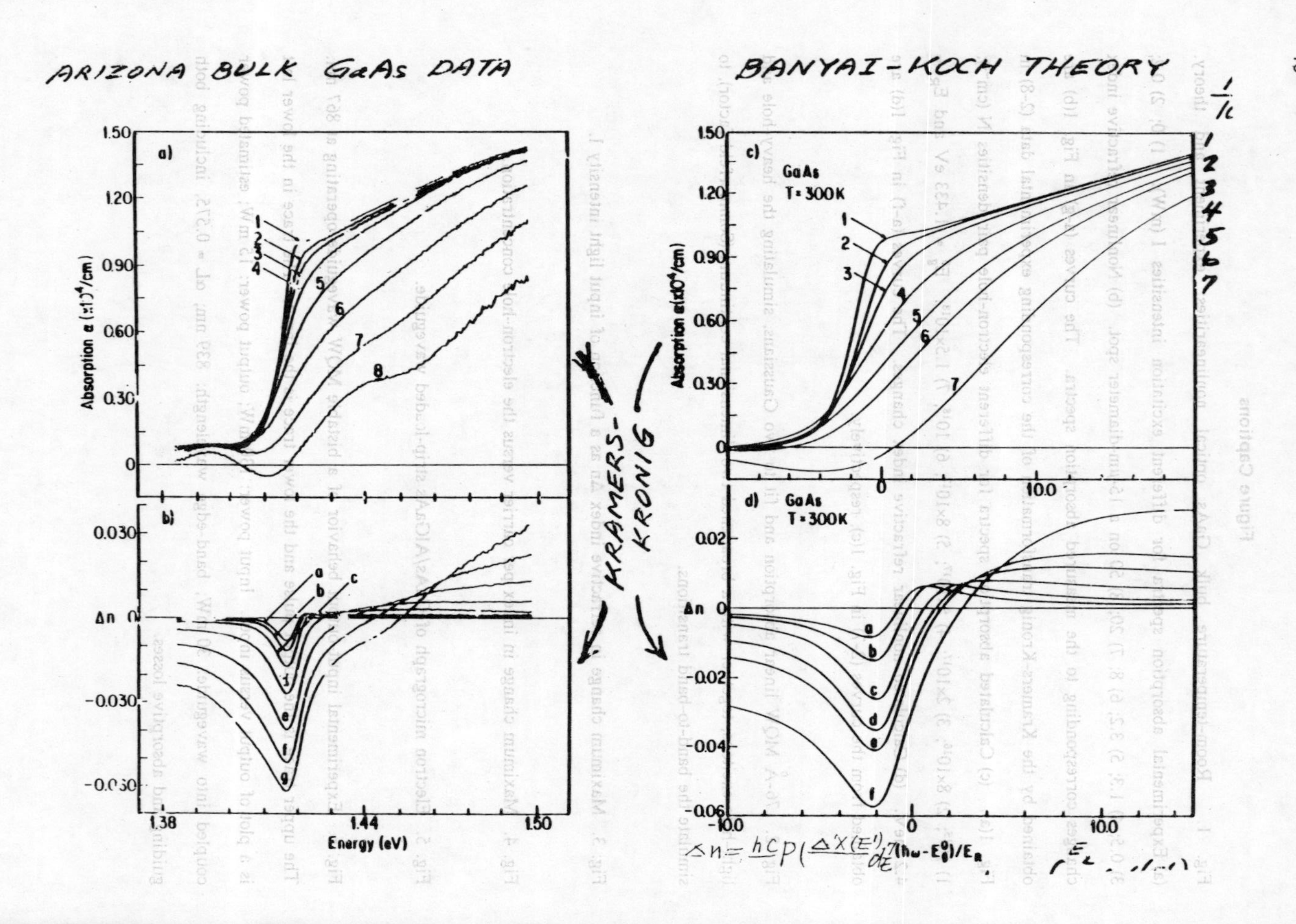
ARIZONA BULK GaAs DATA
BANYAI-KOCH THEORY
KRAMERS-KRONIG
Absorption α (×10⁴/cm)
Energy (eV)
GaAs
T=300K
Δn

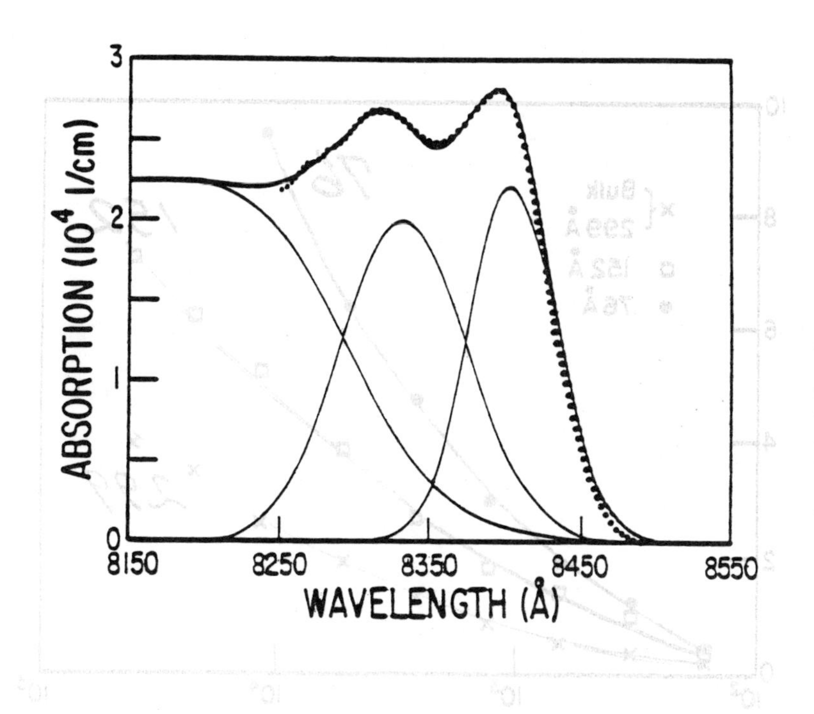

⇒ Δn mostly excitonic in 76 Å

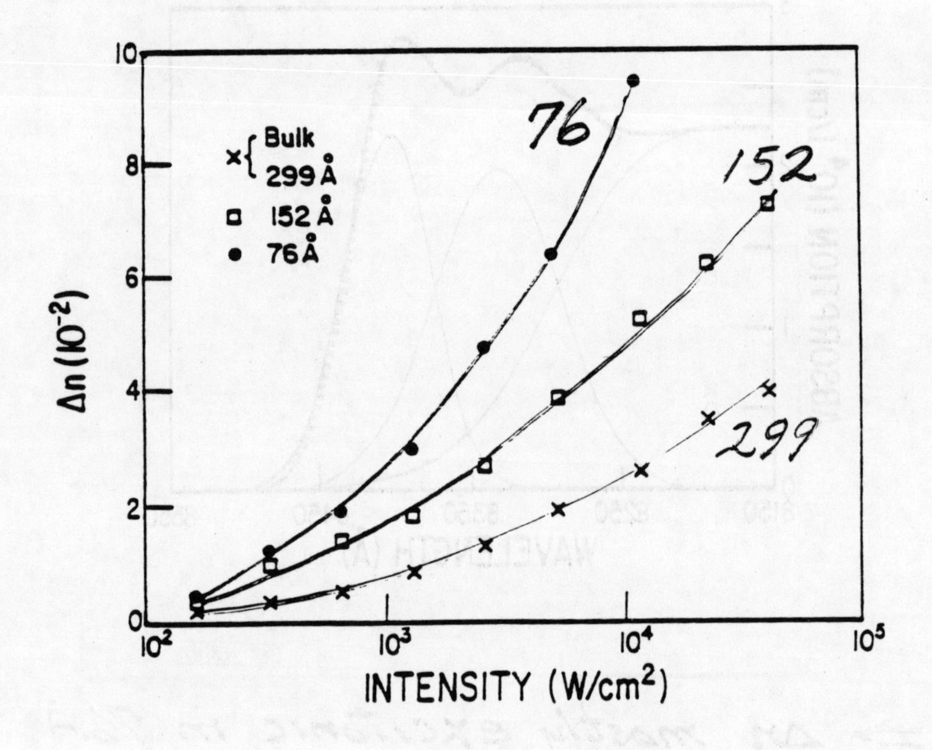
Bulk
299 Å
152 Å
76 Å
76
152
299
Δn (10^-2)
INTENSITY (W/cm^2)

ASSUMES $\tau_{CAR.} = 20$ ns

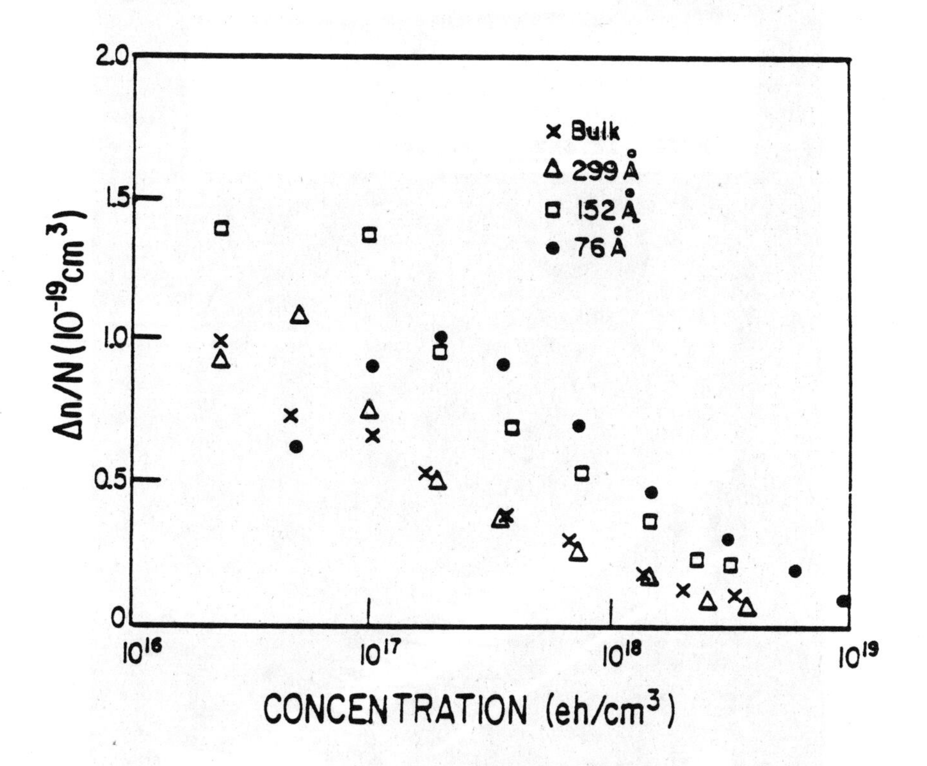

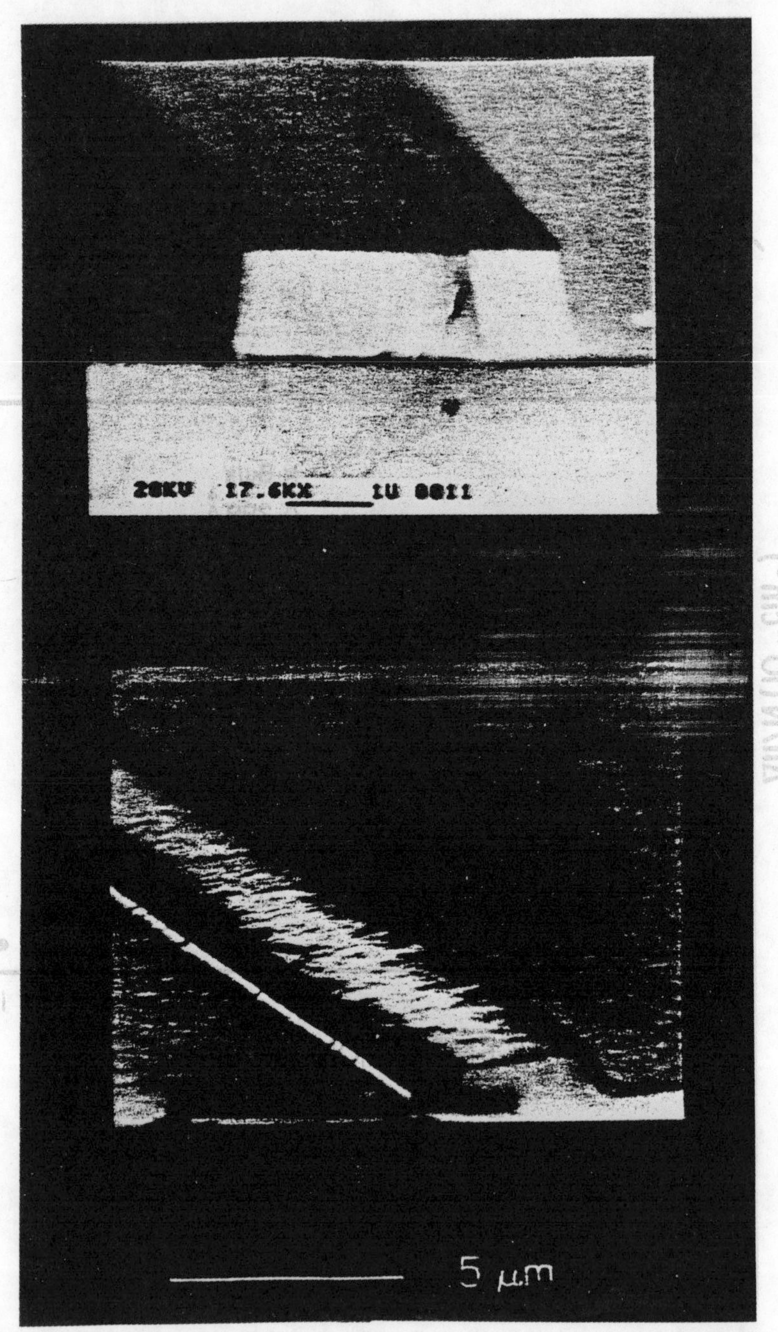
20KV 17.6KX 1U 0011
5 μm

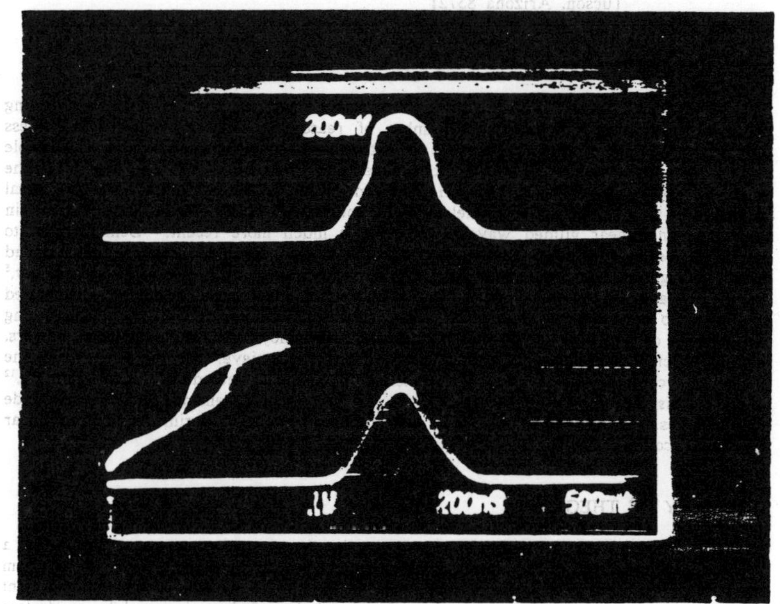
200mV
.1V
200nS
500mV

NONLINEAR DISTRIBUTED COUPLING INTO OPTICAL WAVEGUIDES

G. I. Stegeman, C. T. Seaton, R. M. Fortenberry, R. Moshrefzadeh, and G. Assanto

Optical Sciences Center
University of Arizona
Tucson, Arizona 85721

INTRODUCTION

Thin-film waveguides are a very attractive material geometry for performing nonlinear optics experiments for a number of reasons. They provide diffractionless propagation in one or more dimensions. Furthermore, efficient interactions are possible at low powers becasue of beam confinement to dimensions of the order of the wavelength of light. The application of second-order nonlinear interactions to optical waveguides is a well-developed field.[1,2] In contrast to this, interest and progress in third-order nonlinear guided wave phenomena is much more recent. Demonstrated to date have been degenerate four wave mixing in CS_2-covered[3] and semiconductor-doped glass[4] waveguides, coherent anti-Stokes Raman scattering with monolayer sensitivity,[5] nonlinear waveguiding phenomena,[6,7] and a variety of phenomena related to distributed coupling into nonlinear waveguides.[8-17] One of the disadvantages of distributed coupling via prisms or gratings is decreased coupling efficiency[11,12,16] at high laser powers, making efficient nonlinear interactions more difficult to achieve. On the other hand, the distributed coupling does lead to interesting phenomena including optical limiting,[12] optical bistability,[13,14] switching,[10,12-15,17] and measurement of waveguide nonlinearities.[8-10,12,16] In this paper we summarize our experiments on nonlinear distributed couplers in various time regimes.

SUMMARY OF THEORY

The physics of a nonlinear distributed-input coupler, such as a prism or a diffraction grating, is quite simple. For example, consider light incident inside a prism onto the prism base at an angle θ, such that the parallel component of the incident light wavevector is $n_p k_0 \sin\theta$, where n_p is the refractive index of the prism and $k_0 = \omega/c$ (see Figure 1). Since the transverse (z) dimension over which the incident and guided wave fields interact is of the order of the wavelength, only wavevector conservation parallel to the surfaces is required. Therefore efficient generation of the guided wave is obtained only when the guided-wave wavevector $\beta \cong n_p k_0 \sin\theta$. If the waveguide contains media whose refractive index depends on the local intensity, ($n = n_0 + n_2 I$, where I is the local intensity), then the guided-wave wavevector depends on the guided-wave power, $\beta \rightarrow \beta(P_{gw})$. Therefore, as the guided wave power grows under the coupler, wavevector matching parallel to the surface is lost, leading to a cumulative mismatch in phase between the growing guided wave and the incident field. This results in a decrease in coupling efficiency, an angular shift in the optimum coupling angle, and distortion in the pulse envelope in the case of pulsed excitation.

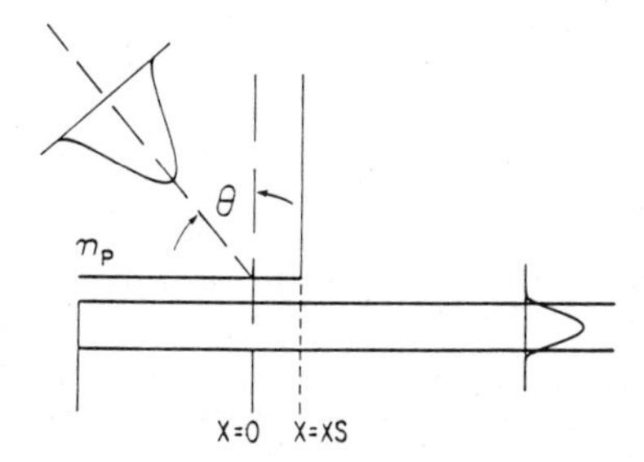

Figure 1. The coupling geometry. A beam, Gaussian along the x-axis, is incident at an angle θ onto the prism base, whose 90° corner is located at xs. A guided wave is generated and propagates for x>xs in the free waveguide.

The mathematical details of the theory of nonlinear distributed coupling can be found elsewhere.[18,19] Writing the field incident as

$$E(\mathbf{r},t) = \frac{1}{2}\hat{e}_{in}a_{in}(x,t)\ \exp[j(\omega t - n_p k_0[\sin\theta x - \cos\theta z])] + c.c. \tag{1}$$

and the guided wave field as

$$E(\mathbf{r},t) = \frac{1}{2}\hat{e}_{gw}a_{gw}(x,t)\ \exp[j(\omega t - \beta(x,t)x)] + c.c. \ , \tag{2}$$

leads to the coupled wave equation[18,19]

$$\frac{d}{dx}a_{gw}(x,t) = \Gamma\ a_{in}(x,t)\ f(z)\ \exp[j\phi(x,t)] - \left[\frac{\alpha}{2} + \frac{1}{\ell}\right]a_{gw}(x,t) \ . \tag{3}$$

Typically the incident light has a field distribution along the base of the prism which is Gaussian, $a_{in}(x) \propto \exp[-x^2/w_0^2]$. The time dependence is either Gaussian with full width at half intensity Δt for pulsed excitation, or a constant for the cw case. For the guided wave field: α is the guided wave attenuation (intensity) coefficient; ℓ is the distance over which a guided wave field amplitude falls to 1/e of its initial value because of reradiation back into the prism alone; $a_{gw}(x,t)$ is the guided wave field amplitude normalized so that $|a_{gw}(x,t)|^2$ is the guided wave power in watts/meter^2; and f(z) is the transverse field distribution.

The nonlinear evolution of the guided wave is governed by the dynamics of $\phi(x,t)$. Neglecting diffusion effects, and including nonlinearities with relaxation times (τ) both faster ($\tau_f >> \Delta t$) and slower ($\Delta t >> \tau_s$) than the pulse width (or power ramping rate), the evolution of the guided wave phase $\phi(x,t)$ is given for the general case by[19]

$$\frac{\partial}{\partial x}\phi(x,t) = \beta_0 - n_p k_0 \sin\theta + k_0\left[A_f|a_{gw}(x,t)|^2 + A_s\left[\frac{1}{\tau_s}\right]\int_{-\infty}^{t} dt'|a_{gw}(x,t')|^2\right] \ . \tag{4}$$

$$A = \frac{\displaystyle\int_{-\infty}^{\infty} dz\ n_2(z)|f(z)|^4}{\displaystyle\int_{-\infty}^{\infty} dz\ |f(z)|^4} \ , \tag{5}$$

where $n_2(z)$ is the appropriate coefficient for the fast (usually Kerr law) or slow (termed "integrating") nonlinearity. From Eq. (4) it is useful to define for integrating nonlinearities

$$n_{2eff} \cong n_2 \frac{\Delta t}{\tau_s}, \tag{6}$$

which shows clearly that the effective nonlinearity for pulses of width Δt is reduced by the ratio of the pulse width to the relaxation time.

These equations can be interpreted as follows. For the instantaneously responding nonlinearity, it is possible to define a nonlinear guided-wave effective index $N(x,t) = N_0 + A_f|a_{gw}(x,t)|^2$, in analogy with an intensity-dependent refractive index $n = n_0 + n_2 I$. Therefore the in-coupled guided wave power reflects instantaneously the power of the incident field and the resulting wavevector mismatch leads to reduced coupling efficiency. For the integrating nonlinearity, the change in the effective index accumulates over the duration of the pulse. Therefore, the effect of the nonlinearity becomes progressively more pronounced, resulting in pulse distortion, as well as reduced coupling efficiency. Thermally induced changes in refractive index subsequent to guided wave absorption are the most common type of integrating nonlinearities.

EXPERIMENTS

To date we have performed nonlinear distributed coupling experiments on five different material systems: liquid-crystal cladding layers,[11] ZnO[12] and ZnS[13] thin-film waveguides, and ion-exchanged semiconductor-doped glass waveguides.[17] In most cases, thermal effects were the dominant nonlinearity encountered, especially at high laser powers. We separate the discussion into experiments on a) integrating nonlinearities in which the pulse width Δt is shorter than the nonlinearity "turn-off" time τ_s, and b) Kerr-like situations in which the optical power is varied slowly relative to the relaxation time τ_f.

Integrating Nonlinearities

Extensive experiments have been performed on thin-film ZnO waveguides. They were fabricated by magnetron sputtering, had propagation losses of the order of dB/cm, were epitaxial with the c axis normal to the substrate surface, and exhibited bulk ZnO refractive indices to ±0.001. Strontium titanate prisms and holographically formed, ion-milled gratings were used for input and output coupling.

Large changes in the coupling efficiency with incident pulse energy were found for both nanosecond (15 ns) and picosecond (25 ps) pulses of 530-nm wavelength (see Figure 2). (The coupling efficiency was first optimized at low powers.) These results are in good agreement with theory, based on either a Kerr law or an integrating nonlinearity, and yield a nonlinearity (effective for the integrating case) with absolute value 2×10^{-16} m^2/W. Nanosecond pulse experiments at 570 nm and 750 nm yielded similar results, showing that the nonlinearity is only weakly wavelength dependent. The peak powers obtained in the picosecond experiments imply that two photon effects should be contributing in the ZnO[20] films, and indeed the difference between the nanosecond and picosecond results is indicative of interference between two photon and absorptive nonlinear mechanisms.

The time response of the nonlinearity in the coupling region was probed by simultaneously in-coupling a 632-nm He-Ne laser beam overlapped in the coupling region with the high-power beam. The recovery time for the coupling of the He-Ne beam was measured to be ≅1 μs, identifying the nonlinearity as integrating with a relaxation time τ_s = ≅1μs. Model calculations for thermal diffusion in the ZnO films predicted a τ of 0.6 μs, implying that the mechanism is thermal.

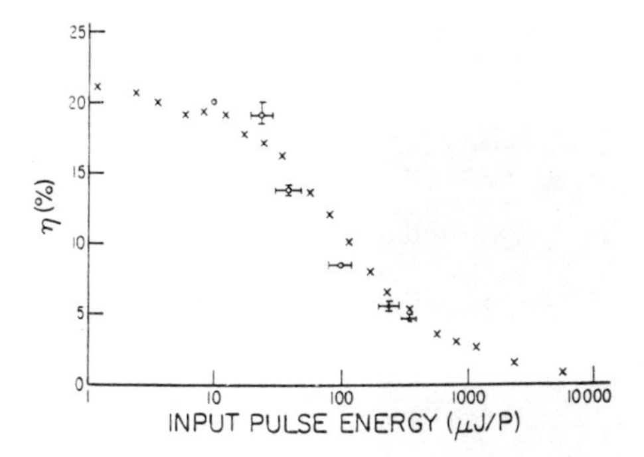

Figure 2. Measured coupling efficiency versus pulse energy at λ = 532 nm for prism coupling into the ZnO waveguides. The x identifies 10 ns pulses and the o the 25 ps pulses.

The variation in the coupling efficiency with incidence angle at low and high powers is shown in Figure 3. Both a broadening of the response curve and a shift in the peak coupling angle are obtained. The peak shift gives $n_{2eff} = +2\times10^{-16}$ m^2/W, which is in agreement with the coupling efficiency measurements. This yields $n_2 = 2\times10^{-14}$ m^2/W for the thermal nonlinearity.

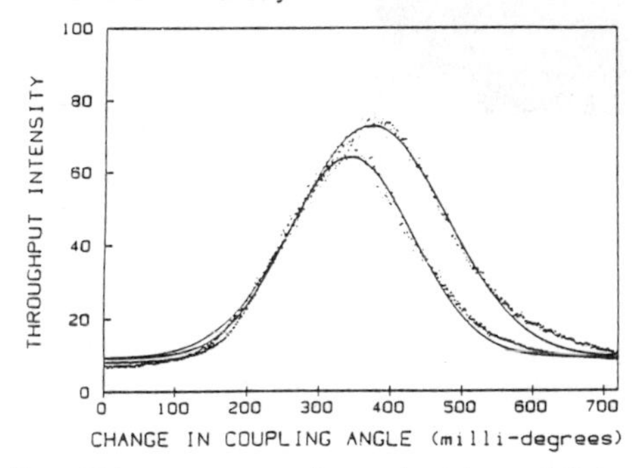

Figure 3. In-coupling efficiency versus coupling angle at low and high powers. The vertical scale of the higher energy pulse (broader one) was adjusted to match the rising edge of the lower energy curve.

Integrating nonlinearities are predicted to produce asymmetric pulse distortion by means of Eq. (3). For a coupler optimized at low powers, the evolution of the temporal envelope of the in-coupled guided wave pulse with increasing energy is shown in Figure 4. Switching up and down within the pulse profile is observed, with switching times which decrease with increasing pulse energy. At high energies, sub-nanosecond switching within the 15-ns incident pulse envelope was measured with a streak camera. At the highest energies used, three maxima were observed in the pulse envelope. Such maxima are indicative of a travelling-wave interaction between the incident field and the guided wave field, with cumulative phase differences between them of multiples of π. The theoretically predicted pulse profiles are also shown in Figure 4, and the agreement with experiment is excellent. Good agreement with theory was also obtained at fixed input energy and variable incidence angle. Although this switching can be fast (sub-nanosecond), the material does not relax for the order of a microsecond. For this time period, a spatial temperature and hence refractive index distribution exists in the coupler region. Therefore, reproducible switching characteristics will not be obtained for pulses whose separation is less than the characteristic relaxation time: thus this phenomenon is not useful for switching high-speed serial data streams.

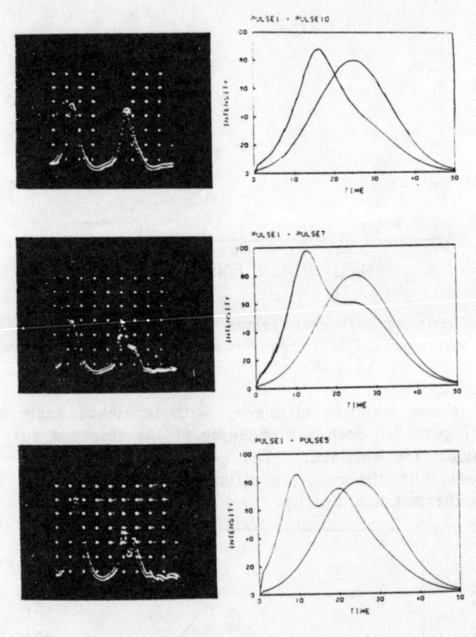

Figure 4. Comparison of experimental and theoretical temporal-pulse profiles at three different energy levels. The first experimental trace in each pair corresponds to the input pulse. Horizontal scale is 10 ns/division.

Grating couplers were also used to verify that the nonlinear coupling process is common to distributed couplers. The coupling grating had a periodicity of 0.34 μm and was ion-milled into the ZnO film to a groove depth of 0.02 μm. The shape of the coupling efficiency versus pulse energy and pulse distortion were qualitatively similar to that obtained with the prism couplers.

Similar phenomena have been observed in coupling to ion-exchanged semiconductor-doped glass waveguides,[21] also with excellent agreement between experiment and theory, for pulse envelopes of hundreds of nanoseconds duration with mode-locked pulses.[17] For single 70-ps pulses,[22] an interference between the very fast electronic and slow thermal nonlinearities was observed in the coupling efficiency experiments (see Figure 5). Furthermore, large changes in the pulse profile were observed when the incident beam was translated relative to the 90° edge of the coupling prism, all of which is in agreement with theory.

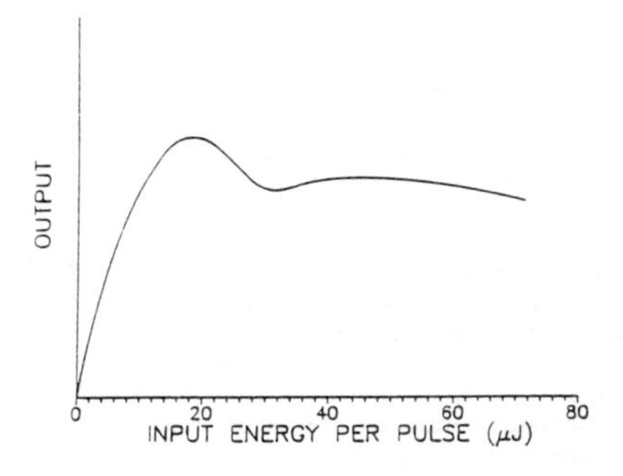

Figure 5. The output versus input energy for prism coupling 100-ps pulses into a semiconductor-doped glass waveguide.

Kerr-Like Nonlinearities

Our most extensive experiments[13] have been performed on ZnS_0 film waveguides which were fabricated by ion-assisted deposition for the first 1000 Å and by normal thermal evaporation for the balance. The power-dependent change in the refractive index has been identified as thermal (via absorption), leading to an intensity-dependent absorption associated with a temperature-induced shift of the band gap toward the excitation wavelength. The angular variation in the coupling efficiency was performed with precise angular positioning and computer-controlled scanning.

The variation in coupling efficiency to the TM_0 mode for increasing incident power (20 mW → 2.6 W) under steady-state conditions is shown in Figure 6. The optimum coupling angle shifts to higher angles and a progressively stronger asymmetry with increasing power occurs on the high angle edge. This edge steepens into a vertical switching characteristic for angular shifts in the peak coupling angle greater than the angular width of the low-power coupling peak. These results are in excellent agreement[13] with the theory outlined here.

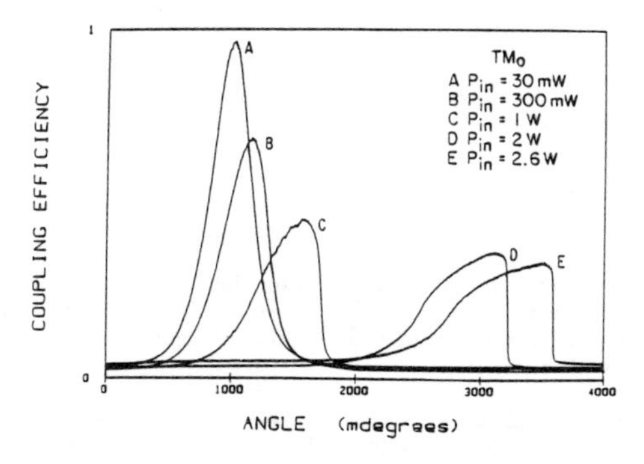

Figure 6. The measured angular variation in coupling efficiency with increasing incident power for a ZnS waveguide.

As the angle for the high-power case is tuned towards the optimum coupling angle, progressively more of the guided wave is coupled in and hence absorbed, raising the film temperature. This in turn increases the effective guided wave index and moves the peak coupling angle (for that temperature) to higher angles. Ultimately incidence angle corresponds to the optimum coupling angle. A further increase in incidence angle produces run-away switch-down because the in-coupled power is reduced, which results in a temperature drop and hence a reduction in optimum coupling angle, which in turn moves the coupling angle further from resonance, resulting in yet less in-coupled power. The net result is switch-down to the low-coupling efficiency associated with the low-power case.

Further experiments at large powers indicated angular bistability as the angle was tuned from above and below the low-power optimum coupling angle. This can be explained either in terms of increasing absorption bistability, or in terms of longitudinal thermal feedback in the coupling region. As expected, power bistability was observed for an incidence angle detuned by 5 linewidths from the low-power case (see Figure 7). No bistability was observed unless the incidence angle was sufficiently detuned to positive angles to intercept a switching characteristic in the angular scans.

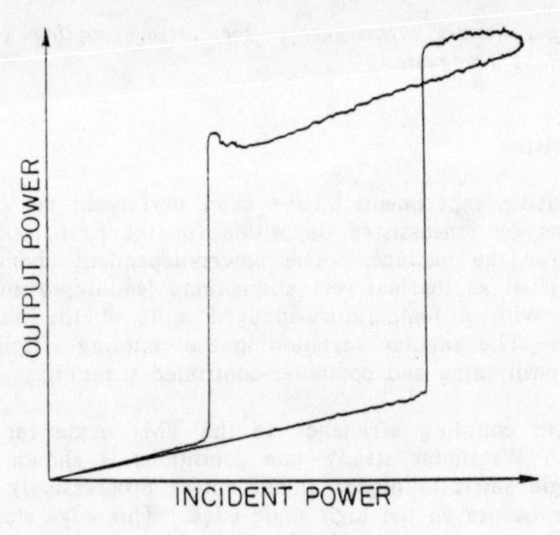

Figure 7. The output versus input energy for both increasing and decreasing power at a detuning of ≅5 linewidths from the optimum low-power coupling angle for a ZnS waveguide.

The reduced coupling efficiency with incident power is an impediment to using optical waveguides for nonlinear optics interactions in nonlinear spectroscopy.[5] As discussed before, the problem is that the effective index N varies with position in the coupler region. However, if for grating coupling the grating wavevector κ (and therefore periodicity) can be varied with distance so that

$$k_0 \sin\theta + \kappa(x) = \beta_0 + A_f |a_{gw}(x)|^2 \, , \tag{7}$$

the synchronous coupling condition can be maintained throughout the coupling region and the coupling efficiency remains optimized. The required form for $\kappa(x)$ is complicated, but it can be approximated by a linear chirp,[18] that is a linear variation in grating periodicity with distance.

This concept was tested with spun-on polystyrene waveguides which exhibit a useful thermal nonlinearity of $n_2 = -10^{-12}$ m^2/W. The procedures outlined in Ref. 23, in conjunction with Reactive-Ion-Beam-Etching, were used to fabricate chirped gratings (chirp rate of $\cong 4\times10^{-3}$ over a 2-mm grating) in a substrate, prior to overcoating with the polystyrene guiding film. The coupling efficiency for the chirped grating, illuminated with a 2-mm-diameter laser beam at $\lambda = 0.515$ μm, was measured as a function of incident power, and the results are shown in Figure 8. In good agreement with theory, the coupling efficiency has a definite optimum which occurs at a well-defined input power. Therefore using chirped gratings it is possible to couple high power beams efficiently into nonlinear waveguides.

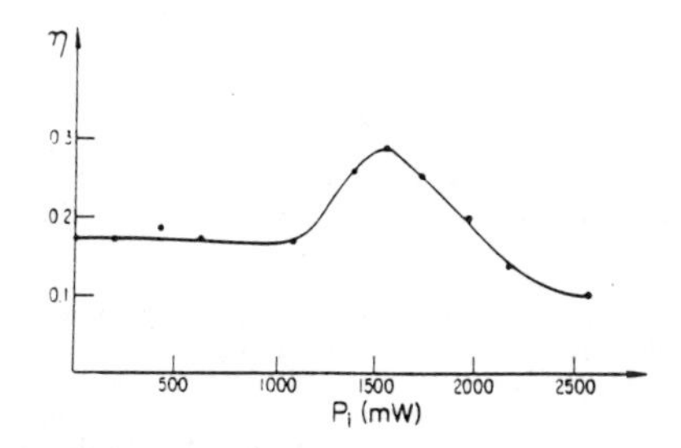

Figure 8. Grating coupling efficiency versus input power for a chirped grating. The nonlinear waveguide was a polystyrene film.

SUMMARY

Prism or grating coupling into integrated optics waveguides is complicated by the presence of waveguide media that exhibit an intensity-dependent refractive index with relaxation ("turn-off") time longer and/or shorter than the rate at which the optical power changes. In all cases, the coupling efficiency decreases with increasing power (for fast nonlinearities) or increasing energy (slow nonlinearities). However, the coupling efficiency can be re-optimized at high powers by using appropriate chirped gratings.

Asymmetric pulse distortion is produced on coupling into waveguides which exhibit nonlinearities with "turn-off" times much longer than the pulse widths. Fast switching and multiple maxima within the pulse envelope are observed because of cumulative phase mismatches in the travelling wave interaction between the incident and guided wave fields.

It has been shown recently[25] that two mechanisms (i.e., nonlocal optically-induced index changes and longitudinal feedback) can lead to bistability when prism coupling a finite cross-section external beam into a nonlinear waveguide.

ACKNOWLEDGMENTS

This research was supported by the Joint Services Optics Program of ARO and AFOSR, the Air Force Office of Scientific Research (AFOSR-84-0277), the NSF/Industry Optical Circuitry Cooperative, and the National Science Foundation (ECS-8304749).

REFERENCES

1. W. Sohler and H. Suche, Integrated Optics III, *Proc. Soc. Photo-Opt. Instr. Eng.* 408:163 (1983).
2. G. I. Stegeman and C. T. Seaton, *Appl. Phys. Rev.* 58:R57 (1985).
3. C. Karaguleff, G. I. Stegeman, R. Zanoni, and C. T. Seaton, *Appl. Phys. Lett.* 7:621 (1985).
4. A. Gabel, K. W. Delong, C. T. Seaton, and G. I. Stegeman, Efficient degenerate four-wave mixing in an ion-exchanged semiconductor-doped glass waveguide, *Appl. Phys. Lett.*, submitted.
5. W. M. Hetherington III, Z. Z. Ho, E. W. Koenig, R. M. Fortenberry, and G. I. Stegeman, *Chem. Phys. Lett.* 128:150 (1986).
6. H. Vach, C. T. Seaton, G. I. Stegeman, and I. C. Khoo, *Opt. Lett.* 9:238 (1984).
7. I. Bennion, M. J. Goodwin, and W. J. Stewart, *Electron. Lett.* 21:41 (1985).
8. Y. J. Chen and G. M. Carter, *Appl. Phys. Lett.* 41:307 (1982).
9. G. M. Carter, Y. J. Chen, and S. K. Tripathy, *Appl. Phys. Lett.* 43:891 (1983).

10. Y. J. Chen, G. M. Carter, G. J. Sonek, and J. M. Ballantyne, *Appl. Phys. Lett.* 48:272 (1986).
11. J. D. Valera, C. T. Seaton, G. I. Stegeman, R. L. Shoemaker, Xu Mai, and C. Liao, *Appl. Phys. Lett.* 45:1013 (1984).
12. R. M. Fortenberry, R. Moshrefzadeh, G. Assanto, Xu Mai, E. M. Wright, C. T. Seaton, and G. I. Stegeman, *Appl. Phys. Lett.* 49:6987 (1986).
13. G. Assanto, B. Svensson, D. Kuchibhatla, U. J. Gibson, C. T. Seaton, and G. I. Stegeman, *Opt. Lett.* 11:644 (1986).
14. W. Lukosz, P. Pirani and V. Briguet, *in:* "Optical Bistability III," H. M. Gibbs, P. Mandel, N. Peyghambarian, and S. D. Smith, eds., Springer-Verlag, Berlin (1986).
15. F. Pardo, A. Koster, H. Chelli, N. Paraire and S. Laval, *in:* "Optical Bistability III," H. M. Gibbs, P. Mandel, N. Peyghambarian, and S. D. Smith, eds., Springer-Verlag, Berlin (1986). F. Pardo, H. Chelli, A. Koster, N. Paraire and S. Laval, *IEEE J. Quant. Electron.*, QE-23:545 (1987).
16. S. Patela, H. Jerominiek, C. Delisle, and R. Tremblay, *J. Appl. Phys.* 60:1591 (1986).
17. G. Assanto, A. Gabel, C. T. Seaton, G. I. Stegeman, C. N. Ironside, and T. J. Cullen, *Electron. Lett.* 23:484 (1987).
18. C. Liao and G. I. Stegeman, *Appl. Phys. Lett.* 44:164 (1984). C. Liao, G. I. Stegeman, C. T. Seaton, R. L. Shoemaker, J. D. Valera, and H. G. Winful, *J. Opt. Soc. Am.* A2:590 (1985)
19. G. Assanto, R. M. Fortenberry, R. Moshrefzadeh, C. T. Seaton, and G. I. Stegeman, *J. Opt. Soc. Am. B*, submitted.
20. E. W. van Stryland, H. Vanherzeele, M. A. Woodall, M. J. Soileau, A. L. Smirl, S. Guha, and T. F. Boggess, *Opt. Eng.* 24:613 (1985).
21. T. J. Cullen, C. N. Ironside, C. T. Seaton, and G. I. Stegeman, *Appl. Phys. Lett.* 49:1403 (1986).
22. G. Assanto, C. T. Seaton, and G.I. Stegeman, unpublished.
23. Xu Mai, R. Moshrefzadeh, U. J. Gibson, G. I. Stegeman, and C. T. Seaton, *Appl. Opt.* 24:3155 (1985).
24. R. Moshrefzadfeh, B. Svensson, Xu Mai, C. T. Seaton, and G. I. Stegeman, "Chirped gratings for efficient coupling into nonlinear waveguides," *Appl. Phys. Lett.*, in press.
25. G. I. Stegeman, G. Assanto, R. Zanoni, C. T. Seaton, E. Garmire, A. A. Maradudin, and R. Reinisch, "Bistability and switching in nonlinear prism coupling," *Appl. Phys. Lett.*, submitted.

Optical Bistabilities with Liquid Crystals

Professor I. C. Khoo
Department of Electrical Engineering
The Pennsylvania State University
University Park, PA 16802

Abstract

As a result of the extremely large orientational and thermal nonlinearities of nematic liquid crystals, various optical bistabilities, switching and wave mixing effects can be realized with low power lasers. We will present a review of recent theories and experiments on some of these new effects.

Optical Bistabilities with Liquid Crystals

Summary

The extraordinarily large orientational and thermal nonlinearities and many other unique physical characteristics of nematic liquid crystals (Ease of fabrication, thin film and integrated optics compatibility, transparency at visible through infrared regimes, ...etc.) have been successfully employed for demonstrating a myriad of interesting nonlinear optical processes.[1] These include optical phase conjunctions, bistabilities, beam amplifications, switchings, image processings and stimulated scatterings. In optical bistabilities, three distinct types of configuration have been demonstrated: Fabry-Perot type, which involves a longitudinal intensity dependent phase shift in a cavity; transverse bistability which involves a single reflection feedback (no cavity) and the transverse phase shift experienced by the laser beam; bistability near the frustrated total internal reflection state. Detailed theories for these three types of bistabilities have been developed and they shed new lights on both the bistability processes and the optical nonlinearities of liquid crystals.

In this paper we will briefly review some of the important fundamental features of these bistabilities processes, particularly the transverse bistability and the frustrated total internal reflection bistabilities. In particular, these theories provide further insights into related nonlinear processes such as optical self-limiting and optical intensity switching and modulations. We will also present recent new experimental results performed on thin film of nematics that show that there are among a handful of existing promising nonlinear materials for applications. For example, using a nematic liquid crystal filled Fabry Perot etalon, we show that fairly fast bistability switching (nanoseconds on-time, microseconds off-times) can be obtained with the thermal nonlinearity. Furthermore, with appropriate dye-doping, the liquid crystal Fabry Perot will switch in the near infrared regions (0.8 μm, 1.3 μm, 1.5 μm) with very low power requirement (microwatts).

For transverse bistability,[2] we will show, with explicit theoretical calculations, that diffractions play a major role in both the qualitative and quantitative aspects of the switching process. The expressions for the conditions governing the existence of bistability and the switching intensities are derived, and they show that in miniaturizing the device, several geometrical-optical parameters cannot be simply extrapolated. The roles of diffusions and other nonlocal processes and saturation effects are also explicitly discussed.

In bistability processes involving the electrodynamics near the total internal reflection state,[3] we explicitly calculated the effects of the (oblique) Fabry Perot action that inevitably accompany the switching from TIR to transmission state, owing to the high reflectivity just above TIR. An estimate is also made of the propagation time of the evanescent field that shows that optimal configurations (for switching energy and speed) require a different configurations than the usual geometry.

1. I.C. Khoo "Nonlinear Optics of Liquid Crystals" in 'Progress in Optics', ed. E. Wolf. Volume XXV. (1988).
2. I.C. Khoo, J.Y. Hou, T.H. Liu, P.Y. Yan, R.R. Michael and G.M. Finn, JOSA B4 886 (1987).
3. I.C. Khoo and J.Y. Hou, J. Opt. Soc. Am. B2, p. 761 (1985).

RESEARCH WORKS ON OPTICAL BISTABILITY AT PEKING UNIVERSITY AND BEIJING NORMAL UNIVERSITY

ZHANG HE-YI

Department of Physics, Peking University

There are several Labs in Peking University and Beijing Normal University working on optical bistabality and nonlinear optics.

The Lab of Optics, Nonlinear Optics and Nonlinear Spectroscopy of Solids. There are several kinds of lasers in these labs, CW Ar^+ laser, dye laser pumped by CW Ar^+ laser, Quantel YAG:Nd^{3+} laser pumped dye laser, He – Ne laser.

LPE Lab for prepare Ⅲ-Ⅴ semiconductor film and bistable semiconductor laser and semiconductor laser.

Vacuum Deposition Lab for prepare different film system and interference filters for optical bistability.

The Lab of nonlinear optics and magneto-optic effect of narrow band gap semiconductors. There are several kinds of infrared lasers in this lab, CW CO_2 laser, CW CO laser, TEA CO_2 laser and tunable CO_2 laser.

Several research works in these labs will be reported briefly.

1. Nonlinear Optical Interface

Nonlinear optical interface(the interface between linear optical medium and nonlinear optical medium) was first proposed by A.Kaplan[1](1977). Then P.W.Smith *et al* have studied the nonlinear interface between glass and CS_2 by using a Q-switch Ruby laser.They also studied the nonlinear interface between LiF and liquid suspension of dielectric spheres as an artificial Kerr medium by using Ar^+ laser[2] . All these results showed the reflection coefficients jumps abruptly but no bistability occurs.

We have studied nonlinear optical interface between glass and liquid crystal(MBBA). The advantage of liquid crystal is its Kerr coefficient n_2 is 100 times larger than the CS_2 for liquid crystal in isotropic liquid phase, when it is in nematic phase, the nonlinear effect is very lasge so it can be convincingly studied by using a Ar^+ laser. For isotropic liquid phase the experimental measurements of reflectivity R as a function of incident intensity for different incident angles are shown in

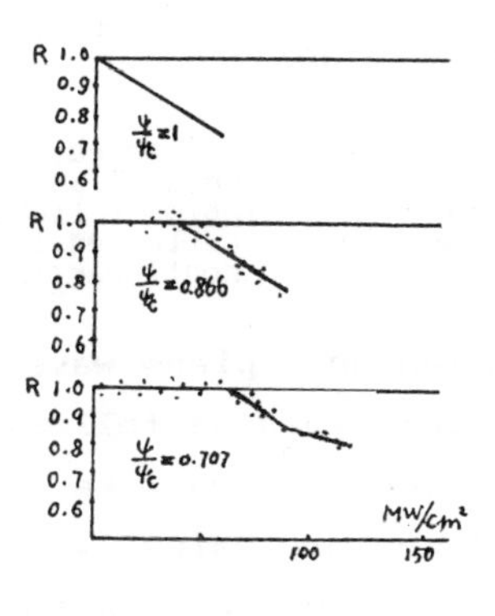

Fig.1 The reflective R of nonlinear interface as a function of light intensity at the interface for different incident angle

According the Snell's law, the total internal reflection angle θ_c is defined by

$$n_0 \sin\theta_c = n_0 - \Delta$$

So

$$n_0 \cos\psi_c = n_0 - \Delta$$

$$n_0 (1 - \psi_c^2/2) = n_0 - \Delta$$

$$\psi_c = \sqrt{2\Delta/n_0}$$

where Δ is the refrative index difference between two medium beside the interface. Δ is a small positive number. The refractive index of nonlinear optical medium is shown as

$$n = n_0 - \Delta + n_2 I$$

where I is the intensity of laser beam in the nonlinear optical medium.

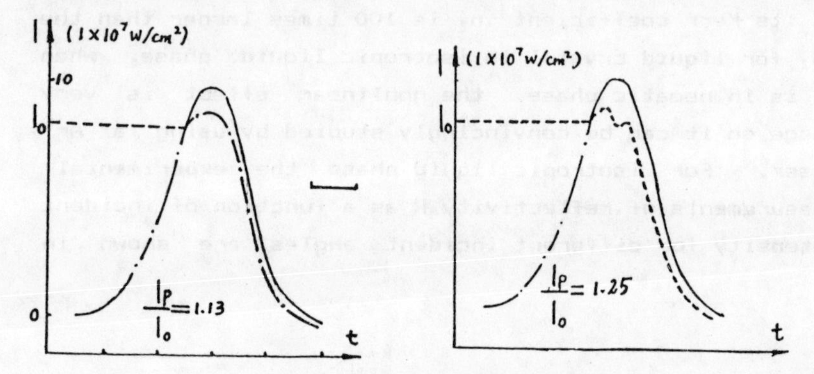

Fig.2 Time resolved reflectivity
I_0 is the threshold, I_p is the peak laser power
Solid line : incident light intensity
Dashed line : reflected light intensity from the nonlinear interface

A.Kaplan's plane wave theory of nonlinear interface can be described as following briefly. For $\psi < \psi_c$, total internal reflection take place. When the light intensity reach threshold value I_0, TIR will change into partial transmission state. The threshold intensity I_0 is determined by following formula

$$n_2 I_0 / \Delta = \begin{cases} -\{1-(\psi/\psi_c)^2\}/2 & 1/\sqrt{2} \leqslant \dfrac{\psi}{\psi_c} < 1 \\ -(\psi_c/\psi)^2/8 & 0 < \dfrac{\psi}{\psi_c} < 1/\sqrt{2} \end{cases}$$

For $I > I_0$ the amplitude reflectivity r as a function of I and ψ/ψ_c is showed as following

$$\frac{n_2 I}{\Delta} = \frac{1}{(1+r)^2}\left[1-\frac{4r}{(1+r)^2}\right]\left(\frac{\psi}{\psi_c}\right)^2$$

For ψ/ψ_c=0.707 the threshold for TIR is

$$I = 8.7\times10^7 \text{ w/cm}^2$$

in which n is given by literature[3]. While the experimental result is $I_0 = 7.8\times10^7$ w/cm²

so the theoretical calculation is in agreement with the experimental result.

Time dependent of reflectivities at the interface was also measured. The results are shown in Fig.2, for $T-T_c=2^{\circ}C$ and $I_p/I_0=1.13$ and $I_p/I_0=1.25$, I_p is the peak value of the incident laser intensity and I_0 is the threshold value. It is apparent that hysteresis occurs in these processes, assuming the laser pulse to be Gaussian in time

$$I(t)=I_p \exp[-(t-t_0)^2/\Delta T^2]$$

the reflected light from the interface is given by

$$I_{rs}(t)=b\int_{t''}^{t} I(t')\exp[-(t-t')/T]dt' \qquad t' \leqslant t$$

T is orientation relaxation time of MBBA (T=250ns). For $T=10\,\Delta T$ calculation leads to results analogous to experimental observations for $I_p=1.3I_0$. This indicates that the hysteresis is attributable to the relatively long MBBA relaxation.

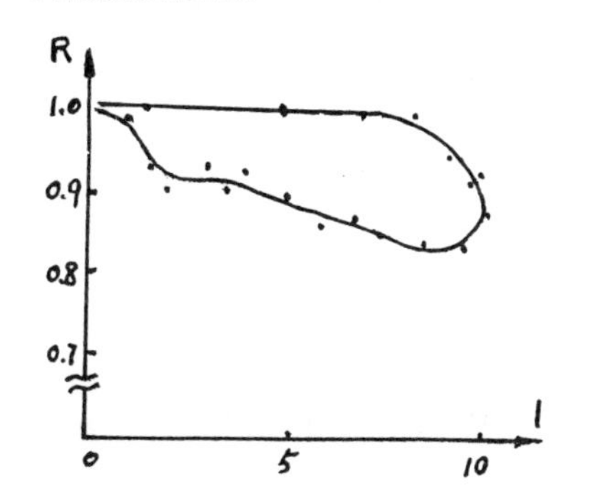

Fig.3 The hysteresis of reflectivity from Fig.2

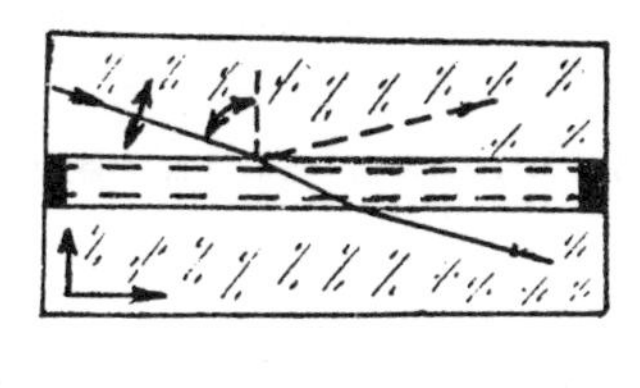

Fig.4 Geometry of liquid crystal cell

The reflection characteristics of MBBA in the nematic mesophase were studied by using an Ar^+ laser. The director $\hat{n}$ of liquid crystal is parellel to the glass surface. The E vector of laser beam is in the incident plane, the index of refraction is given by

$$N(\phi)=N_o N_e (N_o^2 \sin^2\phi + N_e^2 \cos^2\phi)^{-1/2}$$

For $T_c - T = 2\,^\circ C$, we observe a TIR transimission switching at very low incidentt laser powers of 1800W/cm^2 (685mW). The switching phenomena observed are believed to result from absorptive heating of MBBA which causes the index to increase to the MBBA isotropic liquid value of 1.60 [4] a value close to that of the glass. Therefore switching phenomena occurred.

2. Experimental Study of F-P Cavity Optimization in Optical Bistability with Linear Absorption and Thermo-induced Refraction Change

In most of dispersive optical bistable cavity, the absorption of nonlinear optical medium inside the F-P cavity parameters(including R, T of cavity and αD and n_2 of medium) and minimize the switching intensity I_c.

David A.Miller has given an analytical expression for the critical intensity I_c for the critical bistability. He gives an explicit expression of I_c

$$I_c = 1/\beta \times 1/\rho \times 1/\mu_o$$

where $\beta = 3n_2/\lambda\alpha$, n_2 is defined through $n = n_0 + n_2 I$.

$R_\alpha = (1-A)R$, $A = 1 - e^{-\alpha D}$, $F = 4R_\alpha/(1-R_\alpha)^2$, H(F) and G(F) are showed as following

$$H(F) = \{(F+2)[(F+2)^2 + 8F^2]^{1/2} - (F+2)^2 - 2F^2\}^{1/2}$$

$$G(F) = \{3(F+2) - [(F+2)^2 + 8F^2]^{1/2}\}$$

For $R_B = R_F = R$. (R_B and R_F is the intensity reflectivity of the back and the front mirror) the $\rho = 1$. For a given R, μ_o is a function of A. So I_c is a function of A. Consequently,

there exists a suitable A which minimizes I_c.
We use F-P cavity filled with dye solution as the non-linear medium. For dye

$$n = n_0 + n_2 I$$

$$n_2 = \frac{w^2(1-e^{-\alpha D})}{2kD}\frac{dn}{dT} \quad \text{and} \quad \frac{dn}{dT} < 0$$

In our experiment R=0.60, D=0.67mm, we obtain a relationship between I_c and A as shown in Fig.5. The solid curve is Miller's theoretical result, clearly there is qualitative agreement. From experiment we have found A_{opt} =0.24, $A=1-e^{-\alpha D}$. So α =4cm^{-1} for HITC solution C=2 x 10^{-4}mol/L.

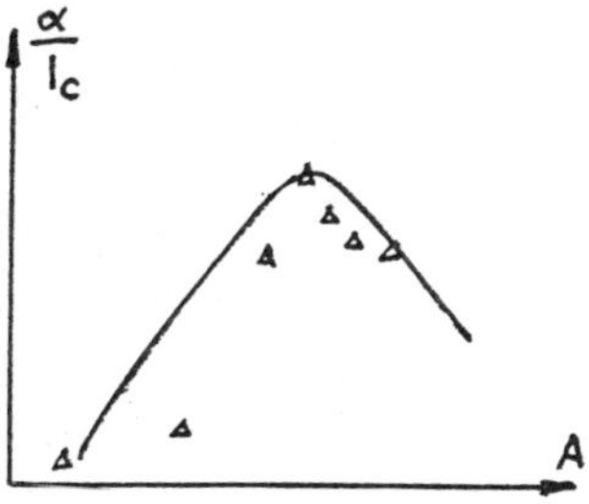

Fig.5 α / I_c as fundtion of absorption $A=1-\exp(-\alpha D)$

'Δ': experimental data
solid line: D.A.Miller's theory

The O.B. loops for different C are shown in Fig.6. There are many orders O.B. loops, the jump height of each order O.B. loop increases as C decreases. Also we can see from the Fig. that input intensity interval between two adjacent order O.B. I increases as absorption decreases. These features can be explained by the following formula

$$T = \frac{(1-R_B)(1-R_A)(1-A)}{(1-R_\alpha)^2} \cdot \frac{1}{1+F\sin(rI_{eff}-\delta_0)}$$

in which T_{max} increases as A decreases. The second feature can also be explained by this formula. For adjacent order the phase shift $\Delta\phi = 2\pi$

$$\phi = 2rI_{eff} - \delta$$

$$\Delta\phi = 2r\,\Delta I_{eff} = 2\pi \qquad r = \frac{12\pi^2 \omega n_2 L}{n_0^2 C^2}$$

$$\Delta I_{eff} = \pi / r$$

for dye solution $n_2 = \dfrac{w^2(1-e^{-\alpha D})}{2KD}\dfrac{dn}{dT}$, $A = 1 - e^{-\alpha D}$

So $\Delta I_{eff} \propto \Delta I_0$ intensity internal two adjacent order OB ΔI_0 increases as A decreases.

3. The Low-power Optical Bistability in Improved ZnS Interference Filters

Recently many scientists use the ZnS, ZnSe interference filter O.B. as an optical logic element. Usually the film system is $G(HL)^m\ 8H(LH)^m$ with m=3, 4. In this OB system the switching on power can be 30 - 50 mW for λ = 5145Å laser with spot size d=50μm to use as a logic gate in optical parallel information processing.

Analysis shows the switching intensity is

$$I_c = \frac{\alpha \lambda K_s f(R_F, R_B, D)}{2\pi\,(\partial n/\partial T) r_0 \cdot \alpha D}$$

In order to decrease switching intensity, we choose the suitable reflect film system and increase the thickness of spacer layer. The system is $G(HL)^6\ 16H(LH)^5$ and $G(HL)^6\ 12H(LH)^4$. Such interference filter O.B. indeed can work at λ=5145Å with switching on power decreases to 5mW.

4. Different kind O.B. of bulk ZnSe slab

We have studied self-focusing of bulk ZnSe plate made by CVD process. The far field pattern was observed systematically. Theoritical analysis shows that it agrees with the experimental results.

We have also studied different kinds of optical bistability of bulk ZnSe slab. Two order transverse optical bistability was observed in bulk ZnSe for first time.

Absorptive optical bistabiliry by using different thickness ZnSe and different laser wave length was also observed.

Negative optical impedance optical bistability was also observed for the first time. The unstable branch $dI_T/dI_I<0$ does occur. The analysis will be discussed in another talk.

5. Nonlinear Optics of Semiconductors and Superlattices

Recently several papers on the nonlinear optics of ZnSe single crystal film and (MnZn)Se superlattices are published. They demonstrate that these thin film have significent third order nonlinearity near the exciton resonant frequences. The estimated lifetime is about 200ps So these thin films are also suitable materials for fast O.B. devices.

A pronounced absorption resonance was evident in the transmission spectra of the ZnSe single crystal film and was associated with the heavy hole exciton. The absorption was observed to be intesity dependent. The linewidth of exciton is about 10Å at 77K. Also (Mn,Zn)Se superlattices was observed. It seems that the spectra of (Mn,Zn)Se superlattices is more complicated than that of ZnSe film.

The nonlinear optical absorption of HgCdTe was also studied systematically.

REFERENCE

[1] A.E.Kaplan Sov. Phys. JETP 45 896(1977)

[2] P.W.Smith *et al*
IEEE J.Q.E. vol.17 pp340-348(1981)
IEEE J.Q.E. vol.20 pp30-36(1984)

[3] G.K.L.Wong and Y.R.Shen
Phys. Rev. A 10 1277(1974)

[4] R.Chang Mol Cryst Liq Cryst 16 53(1972)

[5] D.A.B.Miller 'Optical Bistability and Multistability in the semiconductor InSb' p.115 in Optical Bistability

[6] B.S.Wherrett, A.K.Kar *et al* OPTICA ACTA 33 517(1986)

[7] Y.Hefetz *et al* Appl. Phys. Let. 47 989(1985)

[8] D.R.Andersen *et al* Appl. Phys. Lett. 48 1559(1986)

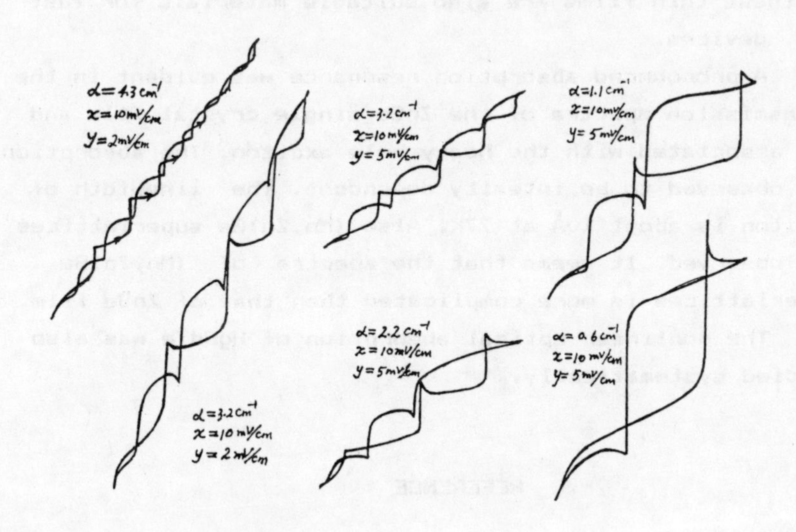

Fig.6 Multi-stage optical bistability in F-P filled with different concentration dye solution

RESEARCH ON OPTICAL BISTABILITY AND OPTICAL COMPUTING IN HARBIN INSTITUTE OF TECHNOLOGY

Chun-fei Li

Department of Physics, Harbin Institute of Technology

Harbin, People's Republic of China

In 1978 our group in Harbin began to work on optical bistability and optical computing, the first group to study these subjects in China. In the last ten years, We have obtained many interesting results which are collected in about 40 papers.

1. Hybrid Optical Bistability

A Hybrid optical bistable device (OBD) consists of an electrooptical modulator and hybrid feedback. Several novel hybrid optical bistable devices were proposed by us in 1979 and 1980, for example:

1) Scanning Non-medium Fabry-Perot OBD[1]

Fig.1 shows a scanning Fabry-Perot with no medium inside. It is a multi-beam interference device.

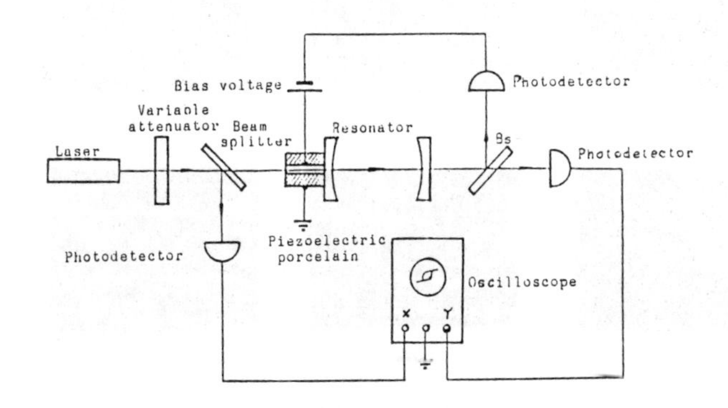

Fig. 1 Scanning Fabry-Perot OBD experiment setup.

The transmitted light is converted into an electronic signal to drive the piezoelectric element which is attached to one of the cavity mirrors. The change of the cavity length causes the phase shift need for optical bistability.

2) Electrooptical Michelson Interferometer OBD[2]

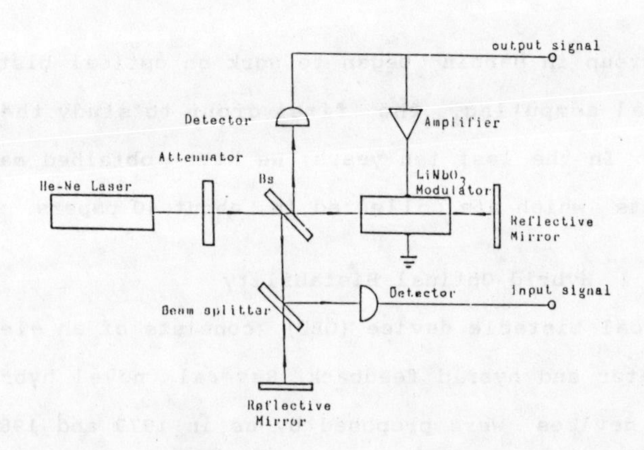

Fig. 2 Electrooptical Michelson interferometer OBD experimental setup.

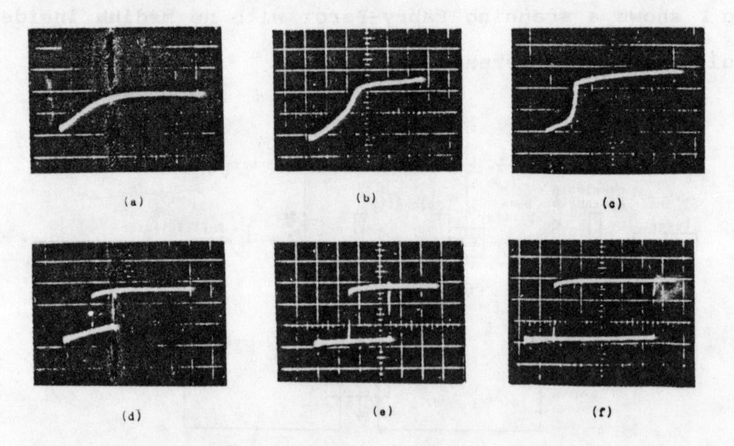

Fig. 3 Optical bistable curves in electrooptical Michelson interferometer.

Fig. 2 shows a two-beam interference device. A $LiNbO_3$ electrooptical modulator is inserted in one of the two arms of a Michelson interferometer. The light transmitted by the interferometer is converted to an electronic signal and fed back to control the modulator.

A group of input-output curves for the different initial phases is shown in Fig. 3.

3) Nonlinear Feedback Electooptical Polarization OBD.[3]

A $LiNbO_3$ modulator with a nonlinear feedback system is placed between two polarizers as shown in Fig. 4. The nonlinear feedback is generated by a silicon controlled rectifier amplifier.

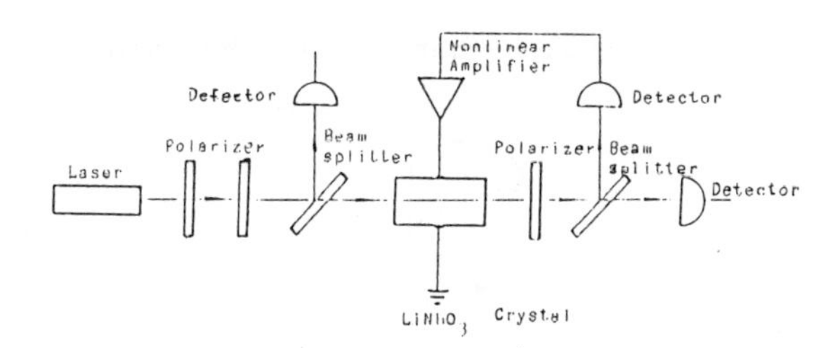

Fig. 4 The electrooptical polarization OBD diagram.

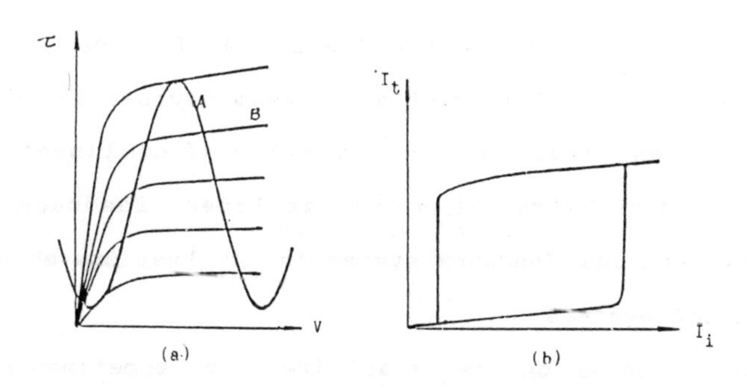

Fig. 5 Optical bistable loop using nonlinear feedback.

We have achieved a better bistable curve with nonlinear feedback than with linear feedback, as shown in Fig. 5.

2. Stability and Instability in Optical Bistability

In 1984-1985 we systematically studied both static stability and dynamic stability in optical bistable devices. Both can be described by a physical quantity, the degree of stability, called S, which is defined as the ratio between relative variations of the input intensity and the output intensity:

1) Static Stability in Optical Bistable Devices.[4]

In the static case, shown in Fig. 6, a bistable curve can be divided into three regions: the stable region where S is greater than 1, the quasi-stable region where S is between 0 and 1, and the unstable region where S is less than 0.

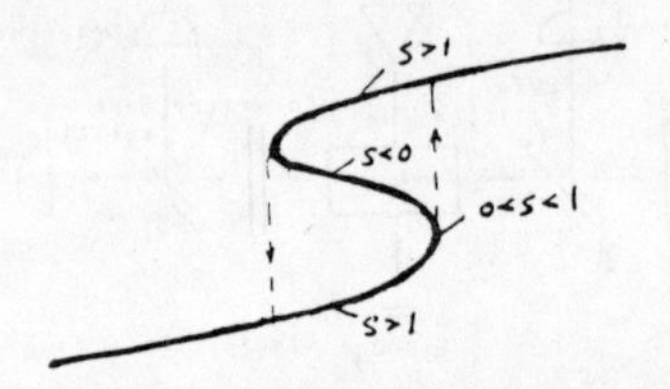

Fig.6 Stability sections on the bistable curve.

According to our theoretical analysis for the S, we conclude that the stability of multi-beam devices is better than that of two-beam devices; the stability of nonlinear feedback systems could be better than that of linear feedback systems; and an exponential feedback system is the best one among nonlinear feedback systems.

We have made an optical stabilizer (or, sometimes called an optical limiter) using a polarization bistable device with an exponential feedback amplifier, its degree of stability is about 50 as shown in the Fig.7.

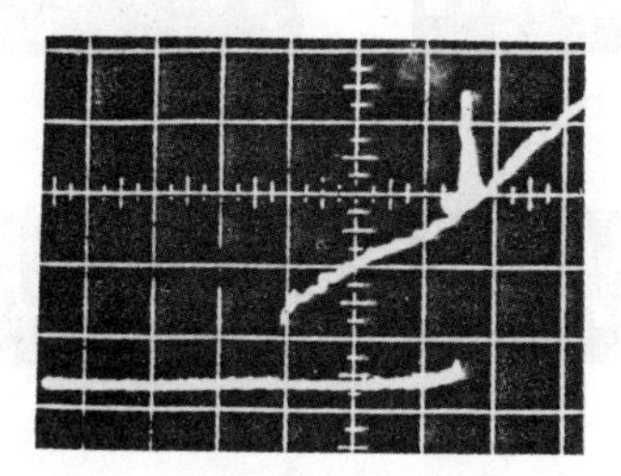

Fig.7 Output vs. input curve for an OBD optical stabilizer with exponential feedback.

2) Instability in OBD with Delayed Feedback.[5]

From the dynamic equations of optical bistability with delayed feedback, using the method of linearized stability analysis, we obtain the threshold condition for instability: S must be greater than or equal to 2. The oscillation period of fundamental unstable mode is 2T for long delay time and 4T for short delay times, where T is the feedback delay time.

In order to test these theoretical results we have used a hybrid OBD with delayed feedback controlled by a microprocessor. The input beam is modulated by an electrooptical modulator.

Fig.8 shows oscillations and chaos in plots of the output intensity verses input intensity for both long and short delay times.

Fig. 9 shows bifurcations and chaos in the plots of output intensity verses the input intensity for both long and short delay times.

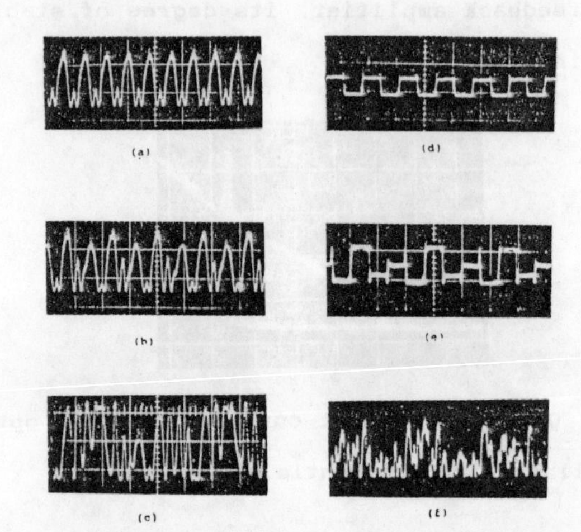

Fig. 8 Output oscillation vs time in a hybrid OBD: (a) period about 4T,(b) Period about 8T and (c) chaos; (d) period 2T, (e) period 4T and (f) chaos.

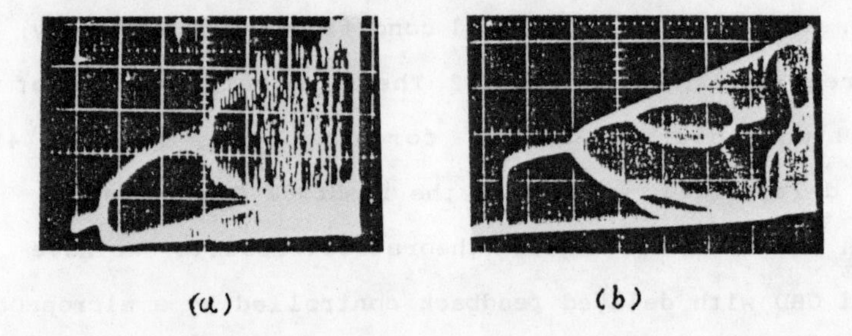

Fig.9 Bifurcations and chaos on the plot of output vs. input for long delay time (a) and short delay time (b).

3.Waveguide Optical Bistability

We have demonstrated optical bistability in both hybrid linear waveguide devices and intrinsic nonlinear waveguide devices

1) Waveguide Electrooptic Phase Modulator OBD[6]

Using a Ti-diffused $LiNbO_3$ waveguide electrooptical phase modulator, we made an OBD on the principle of two-beam interference in 1981. It is similar to the Michelson interferometer OBD, see Fig.10.

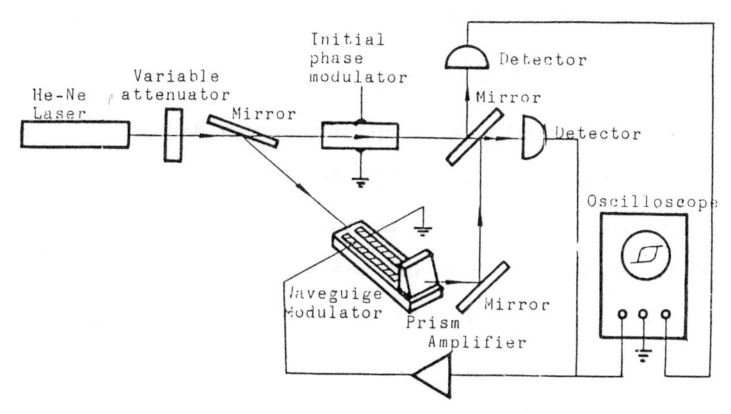

Fig.10 The experiment setup for waveguide electrooptical phase modulator OBD.

2) TE-TM Mode Interference Waveguide OBD.[7]

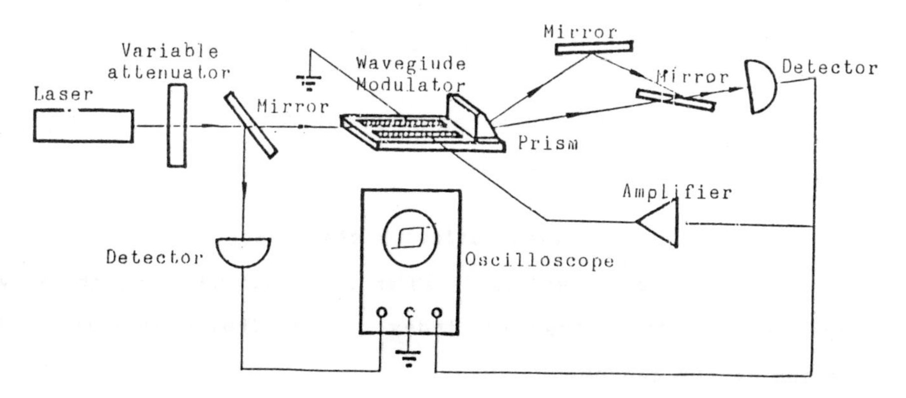

Fig.11 The experiment setup for TE-TM mode interference waveguide OBD.

We use a He-Ne laser to excite the TE mode and the TM mode

in a $LiNbO_3$ waveguide modulator simultaneously. These two beams coming from an output coupling prism are made to interfere with each other, and then the detected signal is fed back to control the modulator shown in Fig.11. Using this device, we achieved optical bistability.

3) Liquid Crystal Nonlinear Coupling Waveguide OBD

Recently we have successfully demonstrated OB in a nonlinear coupling optical waveguide using the liquid crystal MBBA, which is inserted between a prism and the waveguide surface. The experimental setup is shown in Fig.12.

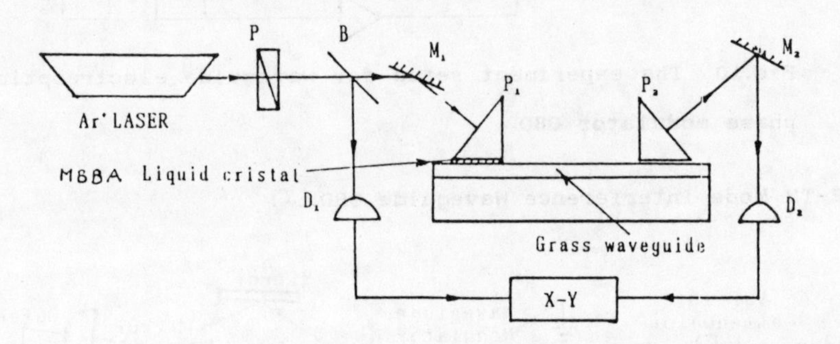

Fig.12 Experiment setup of liquid crystal nonlonear coupling waveguide OBD.

4. Thermal Optical Bistability

Thermal optical bistability is an intrinsic OB, The intensity dependence of the refractive index arises from laser induced heating of the device.

1) Dye-filled Etalons.

Perhaps the simplest device exhibiting thermal optical bistability is the dye-filled etalon, that is, a dye solution between two mirrors. Here we used a cryptocryanine solution. The minimal He-Ne laser power was 4 mW. This work was done in 1982. The experimental photographs are shown in Fig.13.

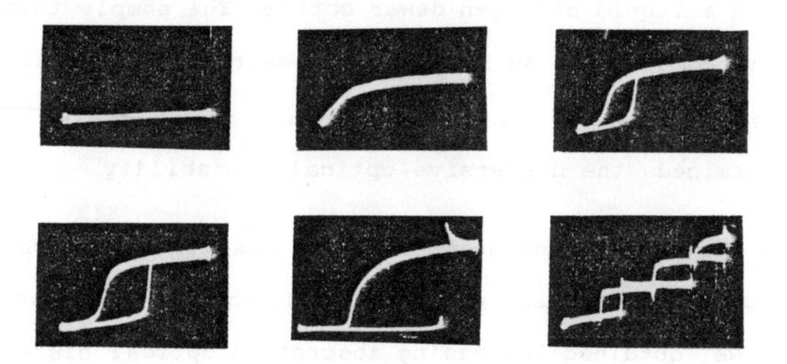

Fig.13 Experimental optical bistable loops in dye-filled etalon.

2) ZnS Interference Filters

Using ZnS interference filters, an argon laser and a rotating plate modulator with a varied metal coating to generate a triangular waveform, we have observed transmissive and reflective bistable loops.

3) Etalons Filled with Organic Adhesive[8]

In 1986 we found, for the first time, a new solid thermal bistable device, this is an etalon filled with Canadian balsam which is easily made. Its switching time is about one millisecond.

5. Semiconductor Optical Bistability

In 1984 to 1986, we concentrated on optical bistabilty in narrow gap semiconductors, such as InSb and HgCdTe. Their nonlinearities are due to band filling effect or thermal effects.

1) InSb Dispersive and Increasing Absorption Optical Bistability[9]

An InSb etalon sample with a thickness of 320 micrometers is placed in a liquid nitrogen dewar bottle. The sample temperature is about 85 Kelvin. An incident CO laser beam is modulated with a triangular waveform by a rotating metal-coated CaF_2 plate. We obtained the dispersive optical bistability.

Because the cavity finesse of our InSb sample and the thermal heat sinking conditions are not very good, when we increased the laser power we obtained increasing absorption optical bistability. In our experiment the minimal input power for generating optical bistability was 2.2 kW/cm^2.

B.S.Wherrett and coworkers predicted that the combined effects of both dispersive and increasing absorption bistability might be observed in InSb. This phenomenon was first observed experimentally by our group. The experimental results are shown in Fig.14.

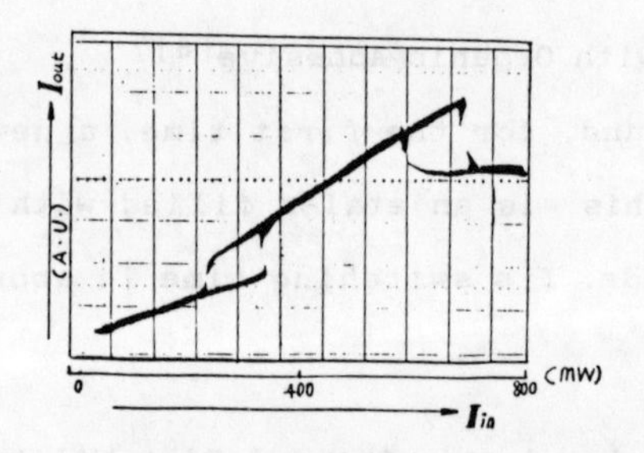

Fig.14 Experimental curve for the combined effects of dispersive and increasing absorption OB in InSb.

2) Room Temperature HgCdTe Two-Photon Optical Bistability[10]

Using a frenquency tunable TEA CO_2 laser with a 100 ns pulse duration, we have obtained optical bistability in a room temperature HgCdTe etalon caused by two-photon excitation. The sample composition coefficient are 0.223 and 0.234. Fig.15 is the experimental results for the the different incident peak power.

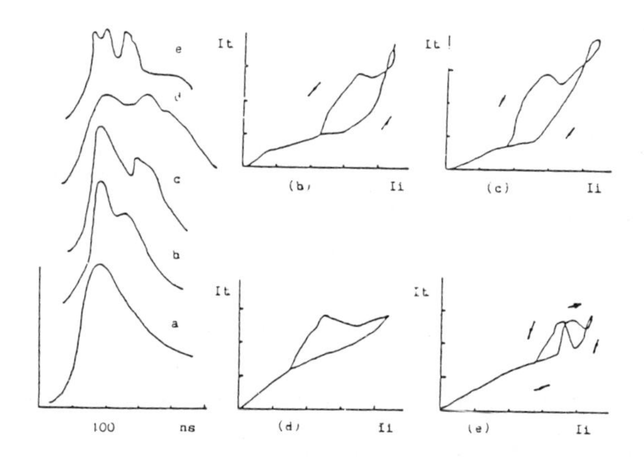

Fig. 15 The experimental plots of HgCdTe etalon with different incident pulse peak power.

3) Thermal Optical Bistability in HgCdTe and Ge[11]

When the laser wavelength is far from the semiconductor band gap, thermally induced optical bistability can still be obtained if the absorption coefficient is big enough. We were the first to demonstrate thermal optical bistability in a $Hg_{0.8}Cd_{0.2}Te$ etalon and in bulk Ge in the case of excitation far from the band gap.

6. Optical Logic Using OBD

In 1984 to 1986 we used dye-filled etalons, thin film interference filters, and light emitting diode bistable devices to study logic functions.

1) Optical Logic in a Dye-filled Nonlinear Etalon[12]

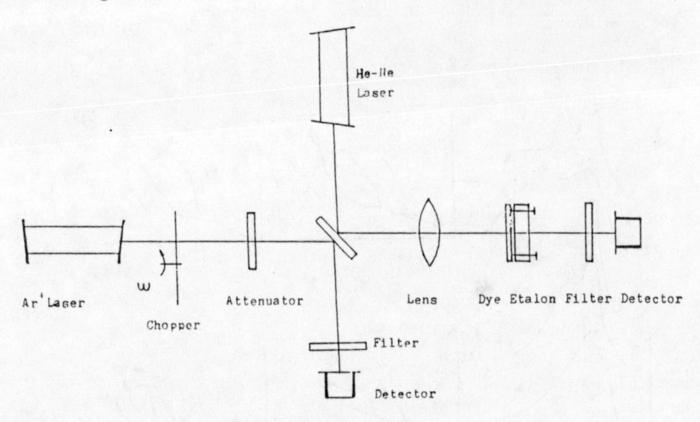

Fig. 16 Experiment setup for demonstration of optical logic using a dye filled etalon.

In Fig.16, an argon ion laser beam passes through a chopper and becomes a pulse of signal light, then goes into dye-filled etalon. A CW He-Ne laser beam is used as the probe light. Changing the initial phase shift, we have obtained five kinds of full-optical logic operations: And, Or, Nand, Nor and Xor shown in Fig.17(a), which are in agreement with our simulations as shown in Fig.17 (b).

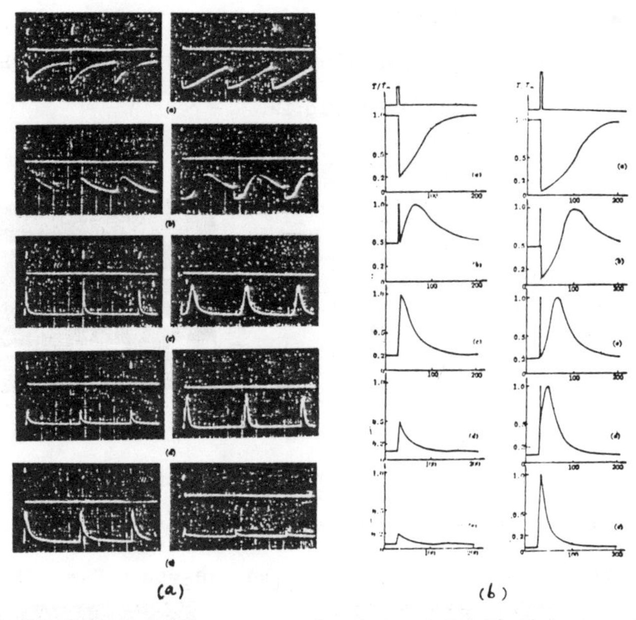

Fig.17 Input and output characteristics vs. time for the dye-filled etalon and simulations. (a) Nor, (b) Nand, (c) Xor, (d) Or and (e) And.

2) Optical Logic in Composite BILED/BILD Circuits[13]

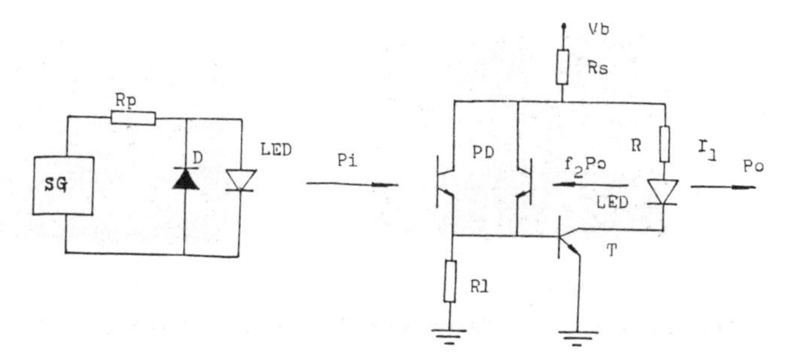

Fig.18 Experimental schematic diagram of BILED.

An optical bistable device which consists of one or two light emitting diodes, a photo-transistor and a triode is called a BILED, as shown in Fig.18.

If an LED branch is connected in parallel with a BILED, one gets a optical invertor, see Fig.19.

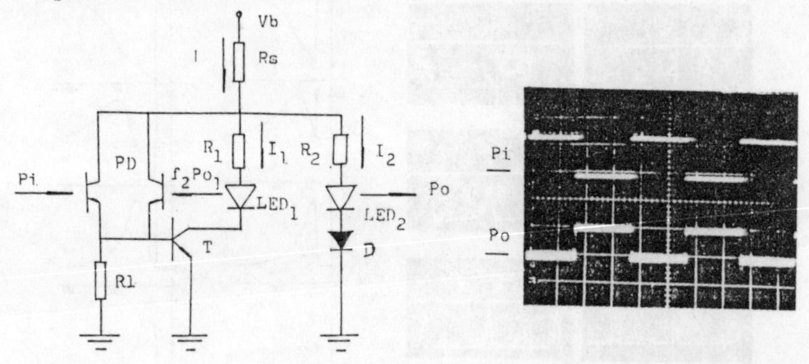

Fig.19 Experimental schematic diagram and characteristicsof optical invertor for a square wave input, where V_b=4.0 V, R_s=20, R_i=0, R_2=50, R_1=110K.

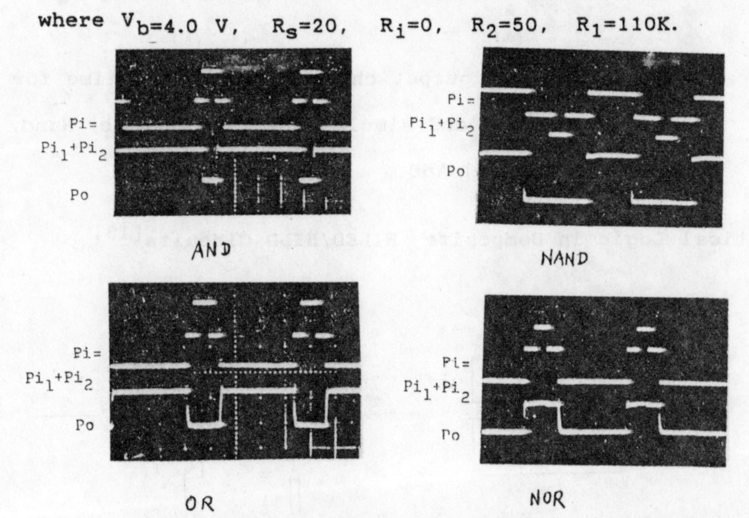

Fig.20 And, Or, Nand and Nor logic gates in BILED.

When the two signal lights P_{i1} and P_{i2} are applied to a BILED simultaneously, one can get the logic of And and Or; and if the signals are applied to a BILED invertor, then Nand and Nor are obtained, see Fig.20.

When two signal lights are applied to two BILEDs connected in parallel, we obtain an Xor logic gate shown in Fig.21.

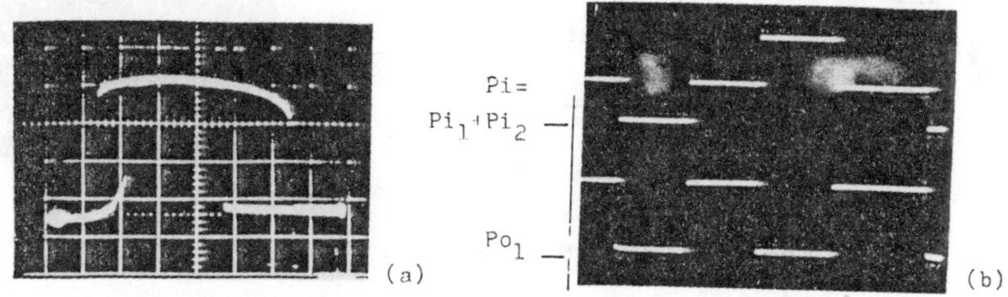

Fig.21 Xor logic when two square wave signals P_{i1} and P_{i2} simultaneously input two BILEDs connected in parallel.

Connecting two BILEDs with each other in parallel, one can obtain the tristable device. If three BILEDs are connected in parallel, one has the fourstable device shown in Fig.22. So in this way, one can get N-stable device using N-1 BILEDs connected in parallel.

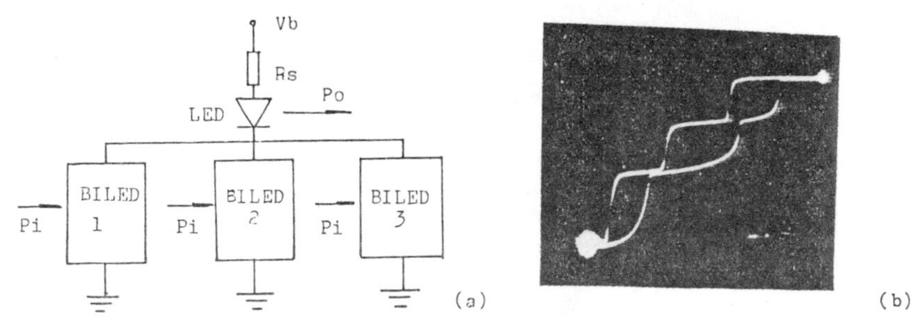

Fig.22. Fourstable device using three BILEDs connected in parallel.

With two parallel compositive BILEDs in the form like this, we get an optical R-S flip-flop shown in Fig.23.

These devices can be built in optoelectronic monolithic integrated circuits for application in optical communication and optical signal processing.

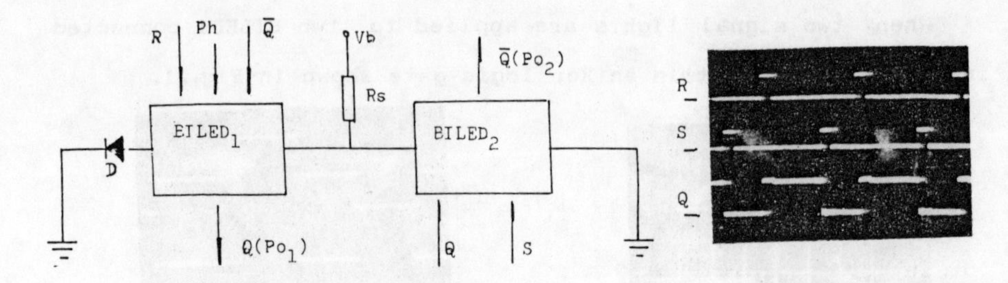

Fig.23 The optical R-S flip-flop when the square wave sig-nals R, S and holding power P_n input a device combined by two BILEDs.

7. Two Dimensional Digital OPtical Computing

Recently we have been interested in two dimensional all-optical computing with symbolic substitutions and shadowgraphs. See Fig.24.

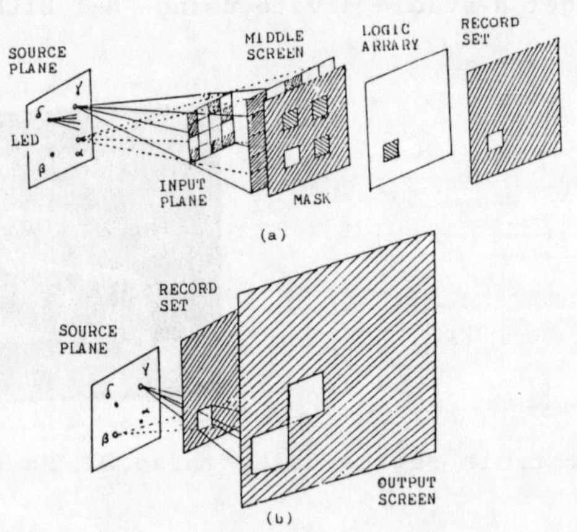

Fig.24 Schematic diagram of implementing symbolic substitution using shadow-casting: (a) recognition phase: (b) substitution phase.

We propose a new optical system for symbolic substitution. The system is composed of a lensless shadow-casting system combined with an optical logical array and optical image storage elements. The numbers and location of the multiple images in symbolic substitution is controlled by an LED array according to the structure of recognition and substitution patterns. This has no lens at all and its structure is very simple, so it is suitable for integrating elements of optical computing.

REFERENCE

1. "Scanning Fabry-Perot interferometer bistable optical device", Li Chunfei and Xu Jing-chun, Chinese Physics, Vol. 2, No. 3, 1982.

2. "Demonstrations of optical bistability using a Michelson interferometer", Chun-fei Li and Jia-ron Ji, IEEE JQE., Vol. QE-17, No. 8, 1981.

3. "Nonlinear feedback bistable optical devices", Li Chunfei, Chinese Physics, Vol. 1, No. 1, 1981.

4. "The static stability and optical stabilizer", Li Chunfei and Chen Lixue, ACTA OPTICA SINICA, Vol.4, No.10, 1984.

5. "Ikeda instability and degree of stability in OB", Chen Lixue, Li Chun-fei and Hong Jing, United States-Japan seminar on conherence, incoherence, and chaos in quantum electronics, Nara, Aug. 30-Sept. 4, 1984.

6. "Study of optical bistability and multistability with optical waveguide modulator", Wan Lide, Jin Enpei and Li Chunfei, J. of Electronics, Vol.1, No.2, 1984.

7. "Optical bistable device using a TE-TM mode interference waveguide modulator", Zou Li-xun, Zhang Lei, Wan Li-de and Chun-

fei Li, Chinese Physic, Vol. 5, No. 1, 1985.

8. "Optical bistability and multistability in nonlonear etalons filled with optical adhesive", Chun-fei Li, Ri-bo Wang, Ping Zhou and Lei Zhang, XV international conference on quantum electronics, Baltimor, April 27-May 1, 1987.

9. "Optical bistability with the combination effects of dispersion and increasing absorption in InSb," Chun-fei Li, Shao-chen Yang and Li-xue Chen, XV international conference on quantum electronics, Baltimor, April 27-May 1, 1987.

10. "Room temperature HgCdTe two-photon optical bistability", Liu Yudong, Sun Wanjun, Jin Enpei and Li Chunfei, to be published in Chinese Journal of Lasers, 1987.

11. "Thermal optical bistability in HgCdTe and Ge", Mao Xianjun, Chen Lixue and Li Chunfei, International Conference on Laser, Xiamen, China, 1987.

12. "Optical logic in a dye-fiiled nonlinear etalon", Zhang Lei and Li Chunfei, ACTA OPTICA SINICA, Vol.5, No.11, 1985.

13. "Optical multistability and optical logic based on a pair of BILED circuits", Chun-fei Li, Shu-tian Liu and Jie Wu, Conference on lasers and electro-optics.

Semiconductor Bistable Laser with Common Cavity Two Sections (CCTS) Structure

Wang Qiming, Wu Ronghan, Lin Shiming, Li Jianmeng,
Zhao Jianhe, Liu Wenxu, Zhang Quansheng

Institute of Semiconductors
Academia Sinica
Beijing, China

In this paper, a brief review is given of some theoretical and experimental achievements on optical bistability in CCTS GaAs/GaAlAs and GaInAsP/InP lasers obtained recent years in the Institute of Semiconductors.

The samples used in experiments are of GaAs/GaAlAs and GaInAsP/InP CCTS lasers striped by either proton bombardment or ridge waveguide. The threshold current is in a range of 50ma-100ma under continuous operation at R.T. The bistability of the laser comes from saturable absorption in the absorption region contained in optical active layer with an intensive F-P feedback effect.

1. Theoretical analysis includs the following folds:

1) Based on the consideration of interband absorption with K-selection rule, starting from the multimode rate equations, systematic calculation by numerical method to give various factors which effect the two important properties of the device, i.e. the width of the bistable current injection region and the switching time of the bistable lasers are illustrated.

2) The theory of instability for nonlinear differential equa--tions is applied to the multimode rate equations. It shows that three conditions must be satisfied to make the system stable, and conditions of bistability and self-sustained pulsation are deduced from small signal approximation. The former corresponds to a persistantly increasing solution, while the latter, to an enhanced oscillation one. Bistability and self-sustained pulsation are both "third order" nonlinear phenomena and closely related to the nonlinearity of gain function. But the self-pulsa-

tion takes place only when the gain coefficient possesses a certain nonlinearity.

3) A Q-switched modulation and gain locked effect existed in the device imply that a stable single longitudinal mode emission can be easily obtained in a proper injection current range.

4) With single mode rate equations, optical amplification factor, transient and stable behaviors are computer simulated under external incident light with different directions to laser axis. An external injection light along the lasing axis is very much beneficial to larger light amplification factor.

2. Experimental results

Some experimental results are given as follows:

1) L-I characteristics under DC and pulse operations.
2) Single longitudinal mode light output at ON state.
3) Switching characteristics.
4) Time-resolved spectrum for investigating the mode stability.
5) Long delay and ultrashort light pulse.
6) Light to light amplification with very low external illumination.

Low-power Optical Bistability in Improved ZnS Interference Filters

Wang Wei-Min, Sun Yin-Guan, Xiong Jun, Jiang Zi-Ping
(Department of Physics, Beijing Normal University)

Guo Ping, Shang Shi-Xuan
(Analytical & Testing Center, Beijing Normal University)

Optical Bistability of ZnS and ZnSe interference filters has recently aroused a great deal of interest. The filters were usually made according to the prescription $G(HL)^m 8H(LH)^m$, where m=3 or 4. The notation $(HL)^m$ implies a quarter-wave of high-index material,H,followed by a quarter-wave of low-index material,L,m times. The region 8H is the high-index spacer layer. If the filters are irradiated by 514.5nm Ar-ion laser beam, the critical switching power for bistability is of a few tens of miliwatts with typical beam spot size of 50μm.[1][2] Using filters, $G(HL)^4 4H(LH)^4$ and $G(HL)^3 8H(LH)^3$, our experimental results were consistant with above-mentioned results. However, the power level must be reduced by one or two orders of magnitude for array processing. In this paper we report 5mW switching power optical bistability in improved ZnS filters.

According to the theoretical analysis, the critical switching irradiance[3] is

$$I = \frac{\lambda \alpha k_s}{2\pi (\partial n/\partial T) r_0} \frac{f(R_F, R_B, \alpha D)}{\alpha D} \qquad (1)$$

where R_F and R_B are the reflectivities for the front and back surfaces of the spacer respectively; α is the linear absorption coefficient of the spacer, D is the thickness of the spacer. The cavity factor $f(R_F, R_B, \alpha D)$ is related to the cavity structure.

From formula (1) it can be seen that the following methods may be used to decrease I. The first is reducing α. Because the wavelength (514.5nm) we used is in transparent region of ZnS, α is decreased to 100/cm. Secondly, we adjust R_F, R_B and αD to keep $f/\alpha D$ at the vicinity of its minimum. The calculation shows that if R_F is equal to R_B, the minimum of $f/\alpha D$ is about 2.6; if R_F is smaller than R_B, it is lower than 2.6. The lowest value is 1.4. It is important to note that even if $R_B - R_F$ is only 0.01, the minimum is larger than 2.6 (see Fig.1).

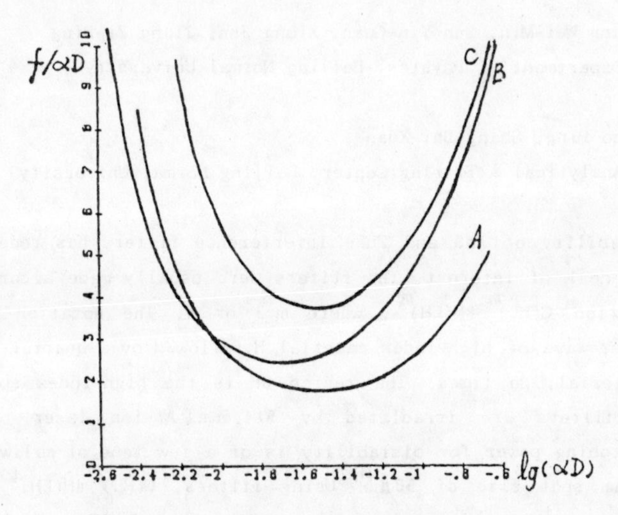

Fig.1

(A) R_F=0.98,R_B=0.99,the minimum is 1.9

(B) R_F=0.99,R_B=0.99,the minimum is 2.6

(C) R_F=0.99,R_B=0.98,the minimum is 3.9

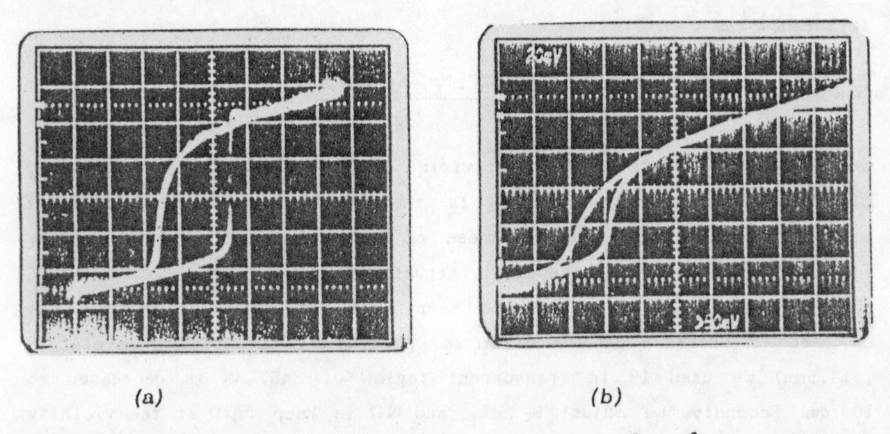

(a) (b)

Fig.2 (a) Sample No.1,P=5.8mw, incidence angle θ=7.6°,the maximum incident power is 10mw. (b)Sample No.2,P=4.5mw, θ=6.6°,the maximum incident power is 15mw.

Thirdly, R_F must be lower than R_B.

In view of above reason, we fabricated supernarrow band filters by thermal-vacuum deposition with a background pressure $8x10^{-6}$ torr. The spacer material we used is ZnS. The reflective stacks are made from a number of pairs of $\lambda/4$ layers of ZnS and Na_3AlF_6. The prescription of our sample No.1 is $G(HL)^6\ 16H(LH)^5$. Its transmission peak wavelength is 518nm. The sample No.2 is $G(HL)^6\ 12H(LH)^4$. The transmission wavelength is 517.3nm. The samples have a full width of 0.8nm at half maximum bandwidth and transmission of 15%. We observed very stable bistable hysterisis loop with maximum incident power of 10mW in the experiment (Fig.2). The switching power P of the sample No.1 and No.2 is 5.8mW and 4.5mW respectively. Using experiment setup introduced by Yasuzi Suzaki[4] we measured the beam spot size. It is 60 microns. This gives I about $175W/cm^2$. The data is lower than early experimental result and in agreement with theoretical analysis.

Reference:

(1) B.S.Wherrett, A.K.Kar et al. OPTICA ACTA. 33, 517 (1986)

(2) Y.T.Chow, B.S.Wherrett et al. J.O.S.A-B 3, 1535 (1986)

(3) B.S.Wherrett, D.Hutching et al. J.O.S.A-B 3, 351 (1986)

(4) Yasuzi Suzaki et al. Appl. Opt. 14, 2809 (1975)

Optical Bistability in Nematic Liquid Crystals due to Laser-Heating Effect

Peng-ye Wang, Hong-jun Zhang, and Jian-hua Dai

Institute of Physics,Academia Sinica,P.O.Box 603,Beijing,China

ABSTRACT

The intrinsic optical bistability due to the laser-heating-induced nematic-isotropic (N-I) phase transition is observed and analysed. The passive optical intensity-limiting and intensity-switching effects are also obtained. The experimental results are in agreement with the theoretical analysis.

Liquid crystal is a kind of nonlinear medium with strong temperature dependence[1]. Consequently, a number of nonlinear optical phenomena can be readily induced by laser-heating effects, such as thermal lens effect[2] and the formation of the isotropic holes in nematic liquid crystals[3,4]. In this paper, we report the intrinsic optical bistability due to the laser-heating-induced nematic-isotropic (N-I) phase transition. Hysteresis and discontinuous changes of the diffraction light are observed. Then the liquid crystal intrinsic optical bistability is performed without the Fabry-Perot interferometer[5] or the feedback mirror[6]. The optical intensity-limiting and intensity-switching effects associated with the N-I phase transition are also obtained.

The experiment was performed at room temperature using E7 which is nematic in the temperature range −10 — 62.5°C. The nematic film was sandwiched between two glass plates coated with transparent conducting electrodes. The thickness of the nematic filem is 15μm. The linearly polarized 5145Å input beam from an argon laser was nearly normally incident on the liquid crystal cell.

The argon laser was operated at TEM_{00} transverse mode. Then the laser beam was Gaussian. The transverse distribution of the intensity can be written as

$$I(r)=I_0 \exp(-2r^2/r_0^2) \quad , \qquad (1)$$

where r is the radial distance, r_0 is the beam radius and I_0 is the central intensity.

Laser-heating can result in temperature rise in the liquid crystal cell. We neglect the heat diffusion in the transverse plane and the temperature distribution in the cell along the direction of the laser beam, because the beam diameter is much larger than the thickness of the cell. Then the transverse distribution of the temperature can be written as

$$T(r)=\alpha I(r)+T_1, \quad (2)$$

where α is a constant and T_1 is the room temperature. We then have

$$T'=T_0 \exp(-2r^2/r_0^2), \quad (3)$$

where $T_0=\alpha I_0$ and $T'=T-T_1$.

As the input intensity increases, the central temperature $T_{max}=T_0+T_1$ will become higher than T_c. The liquid crystal in the central part of the laser beam will transform to the isotropic phase, i.e. a small isotropic hole is formed in the liquid crystal due to laser-heating. The radius r_h of the hole can be obtained from Eq.3,

$$2r_h^2/r_0^2=\ln(T_0/T_c'). \quad (4)$$

The central temperature is equal to T_c or $T_0=T_c'$ at the critical point. Since $T_0=\alpha I_0$ and I_0 is proportional to the input intensity I, the relation between the radius of the isotropic hole and the input intensity can be written as

$$2r_h^2/r_0^2=\ln(I/I_c), \qquad (5)$$

where I_c is the critical threshold intensity.

When the laser beam propagate through the small hole, Fraunhofer diffraction can be observed. A typical far-field diffraction pattern is shown in Fig.1. For higher order diffraction rings, the separation between two neighbouring rings approach the value[7]

$$\Delta\theta = \frac{\lambda}{2r_h} \quad . \qquad (6)$$

This gives

$$r_h = \frac{\lambda}{2\Delta\theta} \qquad (7)$$

We assumed that the temperature is linearly dependent on the input intensity in the above analysis. But the case will be different around the N-I phase transition point. The medium is isotropic in the central part of the laser beam after the phase transition. The small isotropic hole is surrounded by the nematic phase. The boundary is a turbit circular region with the temperature about T_c. This region will have stronger absorption. Therefore, the laser-heating coefficient after the N-I phase transition will be larger than that before the N-I phase transition. Let α_1 and α_2 be the α values before and after the N-I phase transition respectively. Then we will have $\alpha_2 > \alpha_1$ This can be written as

$$\alpha = \alpha_1 + (\alpha_2 - \alpha_1) H(T_0 - T_c') \quad , \tag{8}$$

where H is a Heaviside function.

The sketch of the temperature T_0 versus the beam intensity is shown in Fig.2. The slope of line (I) and line (II) are α_1 and α_2 respectively. If $T_0 < T_c'$, the temperature increases along line (I) as the intensity increases. When T_0 reaches the temperature T_c' $(T_0 + T_1 = T_c)$ the temperature changes discontinuously from line (I) to line (II) because the value of the coefficient α changes from α_1 to α_2. The state of the liquid crystal jumps from A to B in Fig.2. The size of the isotropic hole changes from zero to a finite value r_B. But if the intensity I_0 decreases from a value greater than T_c'/α_1, the temperature will decreases along line (II). The size of the hole assumes a finite value even at $I_0 < T_c'/\alpha_1$. Upon reaching a lower falling threshold intensity $I_0 = T_c'/\alpha_2$, the state changes discontinuously from C to D in Fig.2. The temperature relation jumps back to line (I). We can not find the jumping of the size of the hole at this point because the hole has disappeared at the point C already.

In order to make the diffraction light of the isotropic hole clear, we used a 90° twisted nematic liquid crystal as the sample. A polarizer was placed between the liquid crystal and the observing screen. The direction of the polarizer was parallel to the input laser polarization. The nematic director in the front surface of the cell was perpendicular to the input laser polarization in order to minimize the self-phase modulation. The diffraction light transmitted through the isotropic hole would have a high transmission because the laser polarization was unchanged, and then parallel to the polarizer. Therefore, the diffraction pattern could be observed easily. Beside the isotropic hole, the polarization of the light transmitted through the 90° twisted nematic liquid crystal would be rotated perpendicular to the polarizer. Then the transmission was very weak. The input intensity was changed linearly by an exit modulation signal, and monitored by the output of the monitor signal of the laser. The diffraction patterns were photographed continuously with a cinecamera. The round-trip changing time of the intensity was 25 second. The room temperature was 17.5^{o}C. The size of the light spot on the liquid crystal was about 100μm.

The hysteresis and the discontinuous jump of the state could be readily observed in the experiment. The experimental results are shown in Fig.3. It is seen in Fig.3(a) that the diffraction rings appeared suddenly with increasing light intensity. The size of the isotropic hole was calculated from the separation of rings with Eq.(7). The solid line in Fig.3(b) is the least-squares fit of the experimental data in the decreasing direction of the intensity with Eq.(5), where r_0=61μm and I_C=0.818W. Letting the input intensity to indicate I_0, we obtained α_1=51.87^{o}C/W and α_2=54.99^{o}C/W, and then $\alpha_2 > \alpha_1$. The results are in agreement with the analysis.

In addition, similar to the intensity-limiting effect due to the laser-induced molecular reorientation[8], this device with 90° twisted nematic liquid crystal could possess passive optical intensity-limiting and intensity-switching properties due to the N-I phase transition. If the polarizer was perpendicular to the polarization of the incident beam, the device would have high transmissivity for weak input intensity. But as the intensity increased, the nematic phase would become isotropic phase at a certain intensity. After the N-I phase transition, the laser polarization was not changed by the medium, and then perpendicular to the polarizer as its original. Therefore the device would have low transmissivity for strong incident intensity. Thus the intensity-limiting effect was performed. If we rotated the polarizer by an angle of 90° to make its direction parallel to the laser polarization. The intensity-switching effect could be performed for the same reason. The experimental results are shown in Fig.4. The horizontal and longitudinal axes are the input intensity and the central intensity of the output, respectively.

REFERENCES

1. P. G. de Gennes "The Physics of Liquid Crystals" (Clarendon, Oxford 1974)
2. V.Volterra and E.Wiener-Avnear, Opt. Commun. **12**, 194(1974)
3. V.Volterra and E.Wiener-Avnear, Appl. Phys. **6**, 257(1975)
4. V.F.Kiteava, N.N.Sobolev, A.S.Zolot'ko, Mol. Cryst. Liq. Cryst. **91**, 137(1983)
5. Mi-Mee Cheung, S.D.Durbin, and Y.R.Shen Opt. Lett. **8**, 39(1983)
6. I.C.Khoo, Appl. Phys. Lett. **41**, 909(1982)
   I.C.Khoo, P.Y.Yan, T.H.Liu, S.Shepard and J.Y.Hou, Phys. Rev. **A29**, 2756(1984)
7. M.Born and E.Wolt, "Principles of Optics", (5th ed. Pergamon, Oxford, 1975), Chap.8, P.397
8. I.C.Khoo, G.M.Finn, R.R.Michael, and T.H.Liu, Opt. Lett. **11**, 227(1986)

Figure Captions

Fig.1. A typical Fraunhofer diffraction pattern of the isotropic hole.

Fig.2. Sketch of the discontinuous change and hysteresis of the temperature T_0 versus the input intensity about the N-I phase transition point,

Fig.3. Experimental results of the discontinuous change and hysteresis of (a) the diffraction pattern (b) the radius of the isotropic hole. The arrows indicate the changing direction if the intensity.

Fig.4. Plot of the central output intensity versus the input intensity showing (a) intensity-limiting effect and (b) intensity-switching effect.

SELF FOCUSING AND OPTICAL BISTABILITIES OF BULK ZnSe

Zhang He-yi, Lian Gui-jun, He Xue-hua
Chen Hong-pei, Ai Bin
Physics Department of Peking University

Abstract: Self-focusing and far field ring pattern of ZnSe plate was studied systematically. Theoretical calculation is agreeable with the experimental results. The transverse effect of optical bistability of bulk ZnSe was also studied. Two stage negative logic optical bistability was observed in ZnSe plate for the first time.
The absorptive and dispersive optical bistability of bulk ZnSe was studied by using different wave length as light sourse.

Optical bistability and related nonlinear phenomena have been studied in a number of semiconductors. The mechanism of nonlinear optical effect in different kind of semiconductors are different. Recently the thermally induced refractive index changes in semiconductor gained increasing interest. Optical bistability has been observed in GaAs[1], ZnSe[2], InSb[3]. Here we report the study of self-focusing and far field ring pattern of bulk ZnSe systematically. The study of refractive, absorptive optical bistability and the transverse effect of optical bistability of ZnSe were also discribed here.

1. The self-focusing and far field ring pattern

When a Gaussian beam passes through an absorbing medium, the medium is heated. This may cause refractive index increase (like ZnSe) or refractive index decrease (like dye solution). For both these cases the far field ring pattern can be analyzed by the follwing way.Because the Gaussian beam intensity distribution will produce a ring pattern can be analyzed by the follwing way.Because radially dependent ray-deflection angle distribution, which can be derived from a ray equation and a thermal conductivity equation [4] [5] :

$$\theta(r) = \frac{dn/dT\ [1-\exp(-\alpha l)]\ P\ \{1-\exp[-2r^2/W(\xi)^2]\}}{2\pi K r} + \frac{r}{R(z_0)}$$

$$z_0 < \xi < z_1 \qquad \text{.........(1)}$$

$z_1 - z_0 = l$ is the thickness of the sample, $\theta(r)$ is the deflection angle of the output ray at r, P is the incident laser power, dn/dT is the refractive index temperature coefficient,K is the thermal conductivity, α is the absorption coefficient, W(z) is the beam radius, R(z) is the radius of beam curvature. The first term arises from the nonlinear effect, while the second term is an initial angular dependent of the incident wavefront.

The ZnSe sample was CVD grown, the thickness used in these experiments are 1.1mm and 2.5mm respectively. The TEM_{00} mode Ar^+ laser was used as the light sourse. The maximum output power of λ=5145Å is 2 watts.

We have observed self-focusing of ZnSe plate by using divergent incident beam, the beam waist is located

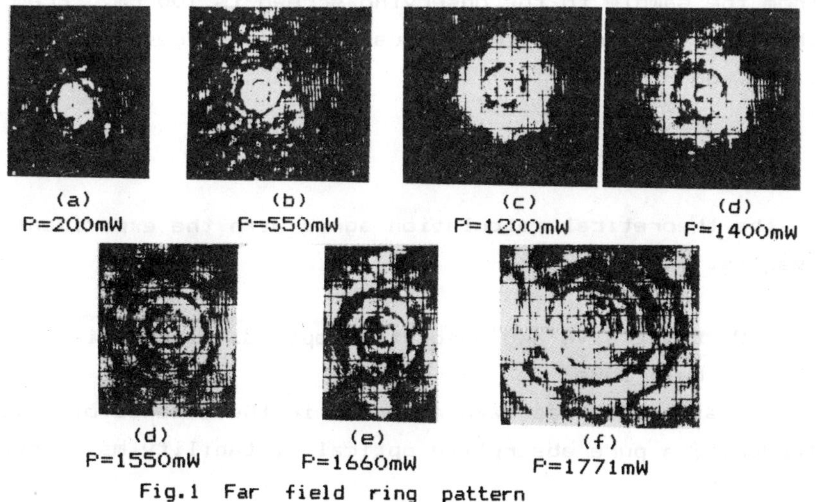

Fig.1 Far field ring pattern

in front of the medium. The far field ring pattern we have observed in a screen was shown in Fig 1. It is found that in the center part there is a bright spot. When the incident laser power was increased, the diameter of the central spot decreased and at the same time the number of rings was increased gradually. We can calculate the diffraction angle of largest ring by using formula (1). For incident laser wave length λ =5145Å, the dn/dT = 2.4×10^{-4} /K for ZnSe while $K = 1.8\times10^{-1}\ W\cdot cm^{-1}\cdot K^{-1}$. In our case, from the beam waist to the front surface of the sample is d=4mm. In this situation, the second term is much smaller than the first term. In the first term, the radius dependent part is

$$\frac{1 - \exp[-(2r^2/W^2)]}{r} \quad \ldots\ldots\ldots(2)$$

this function has a maximum for r/W=0.8, i.e., r=0.8W(z), for beam waist W =20μm, so in the front surface W=38.8μm, with α_0=6cm , P=1.75watts and from formula (1) we get

$$\theta_{max}^{cal} = 3.5$$

The experimental results are shown in Fig.1, the distance from the sample to the observing screen is 100 cm. From the Fig.1 we can get the radius of the largest ring is 6.5cm. So

$$\theta_{max}^{exp} = 3.7$$

So the theoretical calculation agree with the experimental results.

2. Dispersive and absorptive optical bistability of bulk ZnSe

Bistability has been observed in the absence of cavity feedback; a pure absorptive optical bistability may occur if the optical absorption increases as the material becomes more excited. M.R.Taghizadeh et al have studied pure absorptive optical bistability in bulk ZnSe.Recently A.K.Kar and B.S.Wherrett [6] have studied thermal dispersive optical bistability and absorptive bistability of bulk ZnSe in a single experimental run and they give a theoretical analysis of their results. Here we report the dispersive and absorptive optical bistability of bulk ZnSe with different laser wavelength.

In the paper [3] on purely aborptive switching in InSb Wherrett et al showed that for an exponential band edge that decreased linearly with increasing temperature bistability should be observable under two essential conditions

$$\alpha_0 D < 0.18 \qquad I_0 > 2.7T_0/A$$

Here α_0 is the initial absorption coefficient in a sample of length D, I_0 is the incident irradiance level, T_0 the bandedge temperature coefficient and A the thermal constant that determines the temperature rise per absorbed irradiance. They expect that with the cavity feedback on this thermal induced absorptive bistability, both the refractive and absorptive bistability should be observable unambiguously in a single experimental run. The prediction is shown in Fig.2(a).

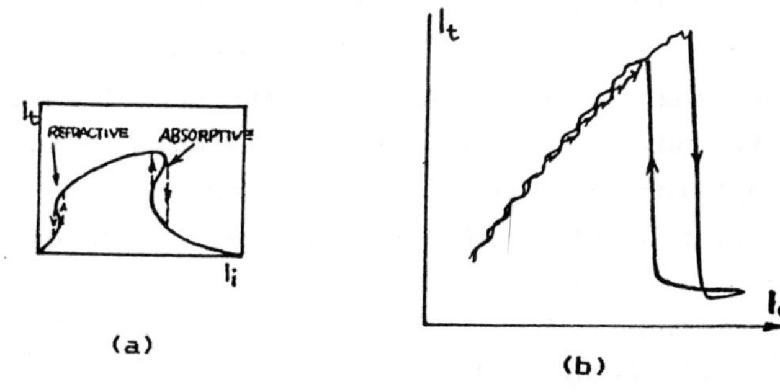

Fig.2 (a) Prediction of consecutive refractive and absorptive optical bistability
(b) Experimental results

The major experimental observation reported in the present paper is shown in Fig.2(b). This confirms the predictions and indeed shows in excess of 11 orders of refractive optical bistability or switching before the onset of dramatic absorption switching.

We have observed dispersive optical bistability of bulk ZnSe. The ZnSe bulk is uncoated F-P cavity O.B. the thickness is 0.8 mm and 1 mm respectively. The two polished surface are paraleled to each other, the imparallelism $\Delta\theta < 1$mrad.

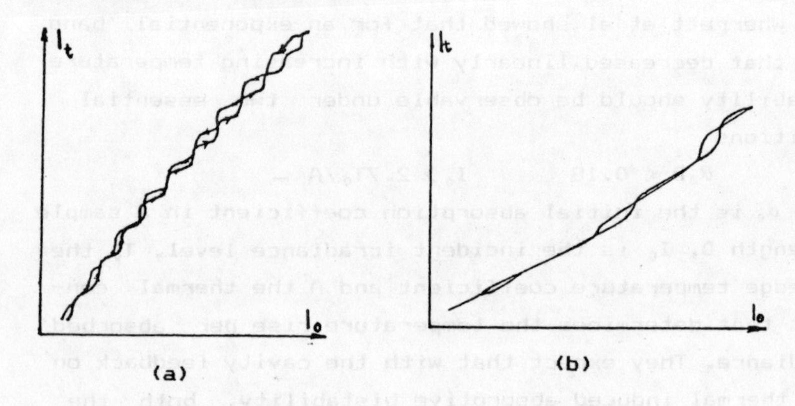

Fig.3 Dispersive optical bistability for different wavelength. (a) =4765A, (b) =5145A

Fig.3(a) and Fig.3(b) are the optical bistability of ZnSe plate O.B. for λ=4765Å and =5145A respectively. It is found that for λ=4765Å the difference of switch-on power of two adjacent order optical bistability is about ΔP=15 mW, but for λ=5145Å $\Delta P \approx 60 - 70$ mW. For two adjacent optical bistable states the optical length difference is

$$\Delta(2nD) = \lambda$$

$$\Delta n = \lambda / 2D$$

and for λ=4765Å, $dn/dT = 4.8\times10^{-4}$/K; λ=5145Å, $dn/dT = 2.4\times10^{-4}$/K.

$$\Delta n = \frac{dn}{dT}\Delta T$$

while $\Delta T \propto P(1-e^{-\alpha D})$

λ =5765Å, α_0 =19.33 cm^{-1}

λ =4880Å, α_0 = 9.88 cm^{-1}

λ =5145Å, α_0 = 6.58 cm^{-1}

The temperature rise for 4765Å, $\Delta T \doteq 0.64$ K

and for 5145Å, $\Delta T \doteq 1.44$ K

$$\frac{\Delta P_{5145}}{\Delta P_{4765}} \doteq 4.6$$

From the experiments $\frac{\Delta P_{5145}}{\Delta P_{4765}} \doteq 4 \sim 4.7$. So it is agree with the calculation.

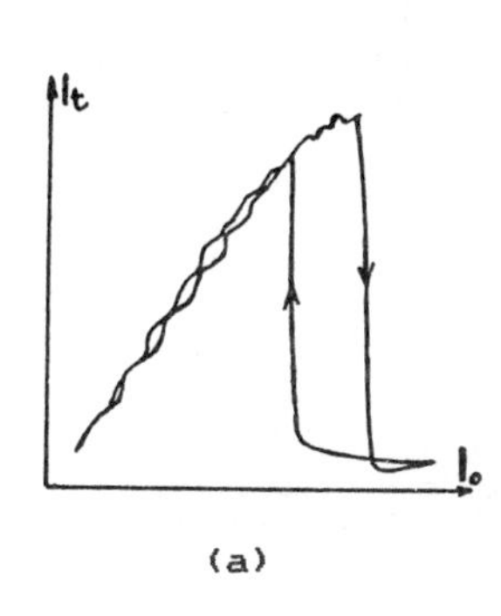

(a)

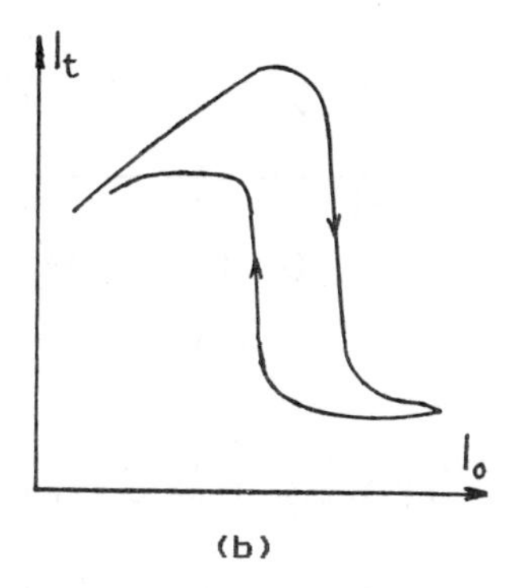

(b)

Fig.4 Absorptive OB in ZnSe at 4765Å (a), 4880Å(b)

Fig.4(a) and 4(b) showed the absorptive optical bi-stability for λ=4765Å and λ=4880Å. In the Fig.4(a) we found the turn down power P_{down} =220 mW for 4765Å and Fig.4(b) showed P_{down} =550 mW for 4880Å. This can analyzed as following

The equations of concern are

$$\alpha = \alpha(\hbar\omega, \Delta T)$$

$$\Delta T = AI_0 [1-\exp(-\alpha D)]$$

$$I_T = BI_0 \exp(-\alpha D)$$

We label $x = \alpha D$, $\ln x = \ln(\alpha) - \ln(D)$

$$\ln[1-\exp(-\alpha D)] = \ln(\Delta T) - \ln(AI_0)$$

Because $\alpha(\Delta T)$ is known empirically for each operating wave length. We can plot $\log(\alpha)$ versus $\log(\Delta T)$ and showed it in Fig.5(a) and (b). From the figures, we found

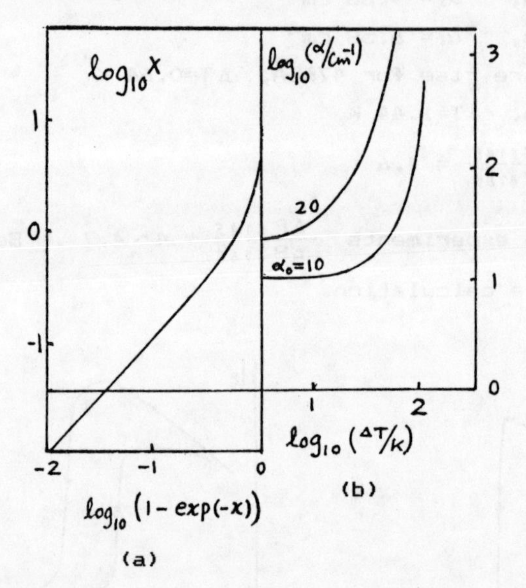

Fig.5 (a) The fixed curve lnx versus ln[1-exp(-x)], to be used as an overlay on (b) in the analysis absorptive bistability.
(b) Temperature dependence of ZnSe absorption for various initial wavelengths

for λ =4765Å the initial α_0=19.33 and temperature rise ΔT is about 70K when the dramatic change of absorption occurs. For λ =4880Å, α_0=9.88, $\Delta T \doteq 100K$.

$$\Delta T \propto P(1-e^{-\alpha D})$$

from initial we roughly estimate $P_{4880}/P_{4765} = 2$. Roughly say it is agree with the experimental results. When we use λ =5145Å no absorptive optical bistability was observed even $P \doteq 2$ watts. It can be analyzed in similar way.

3. The transverse effect of ZnSe plate optical bistability and transverse optical bistability

The sample what we used in this experiment is d= 2.5 mm, when λ =5145Å laser beam was focusing onto the front surface of the sample. The self-focusing was observed. When the incident laser power is increased the spote size is increased too, when the laser power reached at P = 1.26 watts the bright spot changed into a ring pattern suddenly, when laser power is increased to 1.77 watts the diameter of the ring changed discontinuously. When we use a lens to collect the total transmitted laser light and focuse it onto a detector, no optical bistability was observed. When we put an aperture in front of the detector, so only the central part of the transmitted light was detected. The two stages negative logic optical bistability was observed as shown in Fig.6

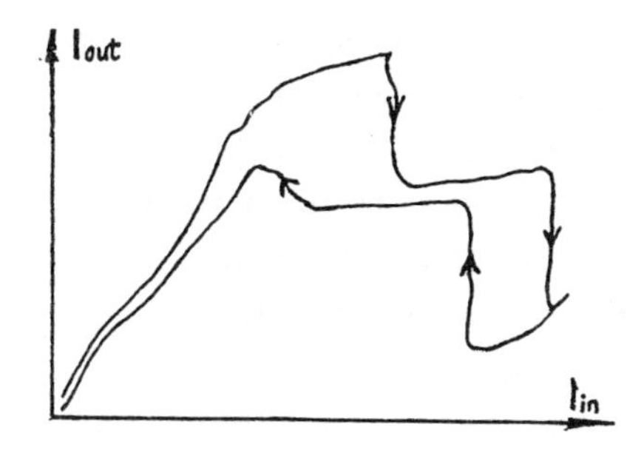

Fig.6 Two stages negative logic OB

The similar experimental results has published for CdS [7]. In that experiment, laser beam was focused into the CdS thin plate when the total transmitted light power was measured, no optical bistability was observed. If a aperture is used to select the central part of the transmitted laser beam, two stage negative logic optical bistability was observed. These phenomena is a kind of transverse effect of optical bistability, and the hysteresis can be also called transverse optical bistability. These kind of optical bistability has been expected by W.J.Firth

[8]. For the low finesse O.B. cavity and Gaussian beam, they found evidence for hysteresis in the profile of the transmitted beam without accompanying power hysteresis. It is called space hysteresis or transverse optical bistability. So our experiments can be analyzed quanlitatively by this theory.

References

[1] H.M.Gibbs et al 'Optical bistability in semiconductors' Appl. Phys. Lett. 35 451-453(1979)

[2] M.R.Taghizadeh, I.Janossy and S.D.Smith 'Optical bistability in bulk ZnSe due to insreasing absorption and self-focusing' Appl. Phys. Lett. 46 331-333 (1985)

[3] B.S.Wherrett, F.A.P.Tooley and S.D.Smith 'Absorption switching and bistability in InSb' Opt. Commun. 52 301-306 (1984)

[4] S.A.Akkmanov, D.P.Krindach, A.V.Migulin, A.P.Sukhorukov and R.V.Khokhlov IEEE J.Quantum Electronics QE-4 568 (1968)

[5] Chun-fei Li, Ping Zhou Optical Bistability p. 260 - 264 (1985)

[6] A.K.Kar and B.S.Wherrett J.O.S.A. B.3 345(1986)

[7] M.Dagenais 'Optical Bistability, Dynamical Nonlinearity and Photonic Logic' pp.75

[8] W.J.Firth and E.M.Wright 'Theory of Gaussian-beam optical bistability' Opt. Commun. 40 233(1982)

BISTABILITY OF THE NONLINEAR PLANAR OPTICAL WAVEGUIDE PRISM COUPLER

Jin Guanghai, ZouPing, Li Chunfei and Li Minying

Department of Physics, Harbin Institute of Technology
Harbin, The People's Republic of China

In the area of nonlinear optics, liquid crystals have been attractive materials since they have one of the largest known intensity - dependent refractive indices. Recently they have proven to be very useful for studying nonlinear effects in waveguides. The existences of nonlinear guided waves [1], limiting action in a prism coupler [2], and bistability in thin-film waveguides with liquid crystal cladding [3] have all been demonstrated. In this paper we report the first observation of bistability in the nonlinear planar optical waveguide prism coupler.

The nonlinear coupler is composed of a planar glass waveguide, a prism and liquid crystal. The gap between the planar waveguide and the prism is filled with liquid crystal whose refractive index is related to light intensity by $n_2 = n_{20} + n_{21} I$.

If a plane wave at the bottom of the prism is represented by $V_p(x) = |V_p(x)| \exp[ikn_p \sin\theta_p x]$, and it is coupled with the transmitting wave $V_F(x) = |V_F(x)| \exp(ik\beta x)$ in the wave guide , then the relation between the input and transmission waves is given by [4]

$$|V_F|^2 = |V_p|^2 \, |t_m|^2 \qquad (1)$$

where t_m is a transfer function. When β is near to β_m^0 which is the propagation constant of guided mode in free waveguide, the relation (1) is shown as

$$|V_F|^2 = \frac{1}{x^2(\beta_m^0)} \cdot \frac{|V_P|}{(n_p \sin\theta_P - \beta_m)^2 + K_m^2} \quad (2)$$

where $x^2(\beta_m^0)$ and K_m^2 are coupling constants. Due to the effect of nonlinear propagation, when optic wave is transported in the wave guide, the propagation constant of guided mode is changed

$$\beta_m = \beta_o' + \Delta\beta_m, \ \Delta\beta_m \ll \beta_o' \quad (3)$$

Applying the perturbation theory, we obtain

$$\beta_o = \beta_o' + a_m |V_F|^2 \quad (4)$$

where $a_m \propto n_{21}$. When the incident angle is dismatching, the initial deturning δ_0 is defined as

$$\delta_0 = n_p \sin\theta_P - \beta_o' \quad (5)$$

Thus, from Eqs (2), (4) and (5), the relation between input and guided waves is obtained

$$|V_F|^2 = \frac{1}{x^2(\beta_o')} \cdot \left[\frac{|V_P|^2}{(\delta_o - a_m |V_F|^2])^2 + K_m^2}\right] \quad (6)$$

$$\text{or} \quad I_F = \frac{c}{x^2(\beta_o')} \cdot \left[\frac{I_P}{(\delta_o - b_m I_F)^2 + K_m^2}\right] \quad (7)$$

Where $b_m \propto a_m$, c is constant.

The experimental results are as follows:

(a) When $\delta_0 = 0$, I_F vs I_P is a monotonous saturation fuction.

(b) When $\delta_0 = 0.001$, the function of I_F vs I_P is an OB loop, and the threshold power is 0.32 W.

(c) When $\delta_0 = -0.001$, the function of I_F vs I_P is another OB loop, and the threshold power is smaller than 0.1 W.

To simplify our discussion, first, we rewrite the relation (7) as ($I_P = I_P(I_F)$), and we consider the differentiation of I_P.

$$dI_P/dI_F = \frac{x^2(\beta_0')}{c} \cdot [K_0' + (\delta_0 - 3b_m I_F)(\delta_0 - b_m I_F)]$$

$$d^2I_P/dI_F^2 = \frac{x^2(\beta_0')}{c} \cdot [-4b\ \delta_0 + 6b_m^2 I_F] \qquad (8)$$

Because dI_P/dI_F is continuous, the condition that knee point of I_P curve exists is as follows: δ_0 and b_m are either both positive, or both negative. In other words, the condition that $I_F \sim I_P$ function appears OB character is that δ_0 and b_m are either both positive or both negative. From the relation

$$b_m \propto n_{21} \propto n_2^{ht} + n_2^{lht},$$

where $n_{21}^{hot} < 0$ and $n_{21}^{Light} > 0$ for MBBA liquid crystal at labouratory temperature, we obtain following conclussions:

(1) When $\delta_0 > 0$, because δ_0 and n_{21}^{light} are both positive, nonlinear optic effect is the main cause of OB loop.

(2) When $\delta_0 < 0$, because δ_0 and n_{21}^{hot} are both negative, thermo—optic effect is the main cause of OB loop.

(3) Because thermo—optic effect is more remarkble than nonlinear optical effect in general, the threshold power of OB when $\delta_0 < 0$ is smaller than the one when $\delta_0 > 0$.

REFERENCES

[1] H. Vach, G. I. Stegeman, C. T. Seaton, and I. C. Khoo, Opt. Lett., 9, 238(1984)

[2] J. D. Valera, C. T. Seaton, G. I. Stegeman, R. L. Shoemaker, Xu Mai, and C. Liao, Appl. Phys. Lett., 45, 1013(1984)

[3] J. D. Valern, B. Svensson, C. T. Seaton, and G. I. Stegeman, Appl. Phys. Lett., 48, 573(1986)

[4] P. K. Tien, and R. Ulrich, J. Opt. Soc. Am., 60, 1325(1970)

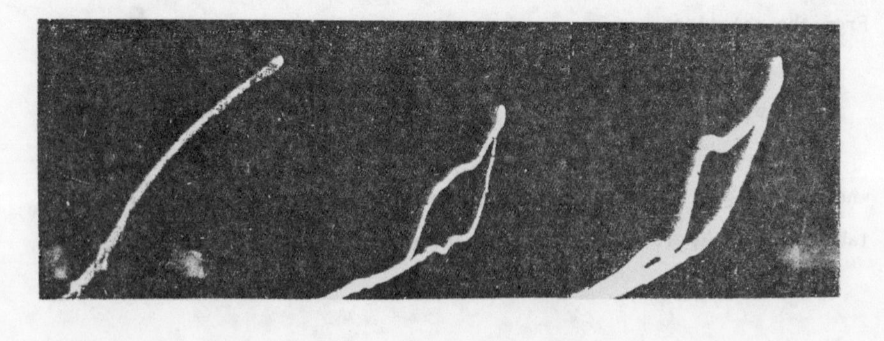

(a) (b) (c)

Fig. Experimental curves of OB for nonlinear coupling optical waveguide using liquid crystal. (a) $\delta_0 = 0$, (b) $\delta_0 = 0.001$, (c) $\delta_0 = -0.001$.

Nonlineaar Three-mirror Fabry-Perot Multistability Element

Guo Ping

Analytical and Testing Center, Beijing Normal University

Sun Yinguan, Xiong Jun, Wang Weiming, Jiang Ziping

Department of Physics , Beijing Normal University

I. Introduction

Since the proposition of nonlinear F-P interferometer(FPI) optical bistable element by Szöke [1],many theoretical and experimental works have been done on such OB element [1-2] . Besides , Szöke proposed three-mirror dual cavities configuration , where the second cavity contains nonlinear medium.This element can be used to produce pulsed outputs for CW inputs . However , the characteristics of two nonlinear coupled cavities has not been analysed yet.In this paper, we deduced the nonlinear three-mirror FPI transmission formula and analysed optical power input-output characteristics of this element using graphic method.

II. Analysis

Three mirrors , which reflectivities are R_1 , R_2 , R_3 respectively , formed dual cavities in which nonlinear medium was filled (Fig.(2.1)).

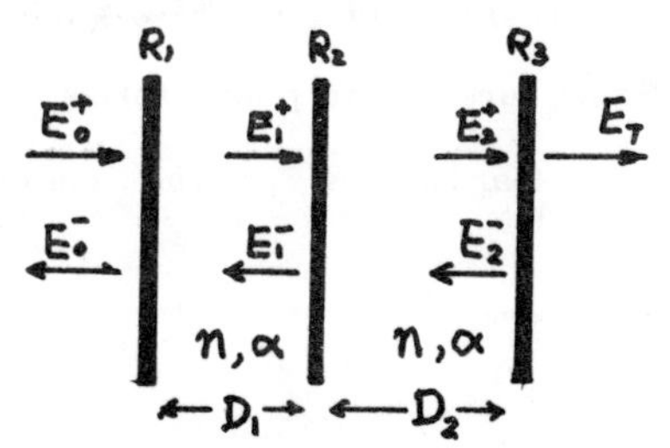

Fig.(2.1) The configuration of nonlinear coupled FPI

The linear absorption coefficient of medium is α, *the refractive index is*

$$n = n_0 + n_2 I , \tag{2.0}$$

where I is intensity.The length of two cavities are D_1, D_2 *respectively. For the sake of convenience , we neglect the mirror losses, assume*

$$R_i + T_i = 1, \qquad i=1,2,3$$

Now we disscuss the transmission of three-mirror nonlinear FPI,our analysis is similar to that of Miller[2],except for three-mirror FPI.

Following Miller's analysis, steady state wave equation is

$$\frac{\partial^2 E}{\partial z^2} + k^2 E = (ik\alpha - \frac{4\pi\omega^2}{c^2}\beta|E|^2)E \tag{2.1}$$

where E is electric field, $P=\beta|E|^2E$ *is nonlinear plarization term. Using plane wave input , assuming*

$$E_i = E_i^+ e^{i\phi_i^+} e^{-ikz} + E_i^- e^{i\phi_i^-} e^{ikz} \tag{2.2}$$

taking slowly varying envelope approximation, we can solve the equation by taking usual boundary condition for a FPI and find E_i

We define the following parameters to characterize the cavities:

$R_\alpha = \sqrt{R_1 R_2}\exp(-\alpha D_1)$, $R_\beta = \sqrt{R_2 R_3}\exp(-\alpha D_2)$, $P_1 = 1-\exp(-\alpha D_1)$, $P_2 = 1-\exp(-\alpha D_2)$; $2\phi_1$, $2\phi_2$, *the first and second cavity round-trip nonlinear phase change respectively;* $k=2\pi n_0/\lambda$, *here* λ *is wavelength in vacuum .* $\delta_1 = kD_1$, $\delta_2 = kD_2$ *are linear phase terms.*

In terms of these parameters , the coupled nonlinear FPI

transmission is given by

$$T = \left|\frac{E_t^+}{E_0^+}\right|^2 = \frac{(1-R_1)(1-R_2)(1-R_3)e^{-\alpha D_1}e^{-\alpha D_2}}{F} \tag{2.3}$$

where $F=\{[1+R_\alpha^2+2R_\alpha\cos(2\phi_1-2\delta_1)][1+R_\beta^2+2R_\beta\cos(2\phi_2-2\delta_2)]$

$+(1-R_2)^2R_1R_3\exp[-2\alpha(D_1+D_2)] + 2(1-R_2)\sqrt{R_1R_3}\exp[-\alpha(D_1+D_2)]\cdot$

$\cdot[R_\alpha R_\beta + R_\alpha\cos(2\phi_2-2\delta_2)+R_\beta\cos(2\phi_1-2\delta_1)+\cos(2\phi_1+2\phi_2-2\delta_1-2\delta_2)]\}$

we find ,

$$\phi_1 = I_0\frac{3\pi n_2}{\alpha\lambda}\frac{(1-R_1)P_1\{(1-R_2)(R_3e^{-2\alpha D_2}-1)e^{-\alpha D_1}+(1+e^{-\alpha D_1})[1+R_\beta^2+2R_\beta\cos(2\phi_2-2\delta_2)]\}}{F} \tag{2.4}$$

$$\phi_2 = I_0\frac{3\pi n_2}{\alpha\lambda}\frac{(1-R_1)(1-R_2)P_2(1+R_3e^{-\alpha D_2})e^{-\alpha D_1}}{F} \tag{2.5}$$

from (2.4) and (2.5) , we can obtain

$$\phi_1 = \phi_2\frac{P_1[(1-R_2)(R_3e^{-\alpha D_2}-1)+(1+e^{\alpha D_1})(1+R_\beta^2+2R_\beta\cos(2\phi_2-2\delta_2)]}{P_2(1-R_2)(1+R_3e^{-\alpha D_2})} \tag{2.6}$$

combining (2.3) and (2.4) gives the second equation for T ,

$$T = \frac{1}{I_0}\frac{\alpha\lambda}{3\pi n_2}\frac{(1-R_3)}{P_2[R_3+e^{\alpha D_2}]} \tag{2.7}$$

In principle,we can solve (2.3), (2.5), (2.6),(2.7) simultaneously to eliminate ϕ_1,ϕ_2, obtain the relation between T and I_0.From $I_T= TI_0$, we can find the input-output **character** function relation.

III. Multistability Graphic Solution

Fig.(3.1) is plotted with the parameters value $\delta_1=\delta_2=19/32\,\pi$, $\alpha=100\text{cm}^{-1}$, $D_1=D_2=0.2\,\mu$, $n_0=2.4$, $R_1=R_3=.94$, $R_2=.96$. It is obviously showing multistability state solution.

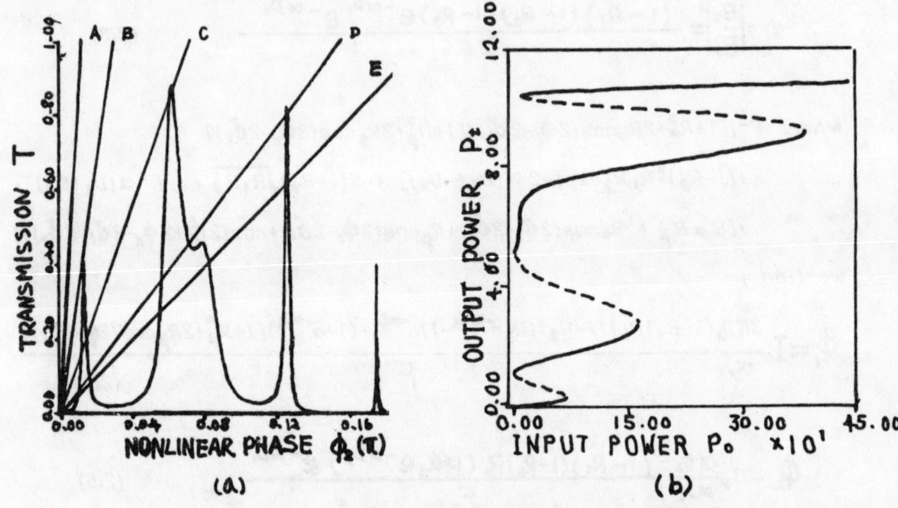

Fig.(3.1) (a) Multistability state graphic solution. The curve is plotted in terms of (2.2), and the straight lines are plotted in terms of (2.7) for different input intensity. A. I_0=1; B. I_0=2.5; C. I_0=6; D. I_0=13; E. I_0=17.

(b) Optical power input-output character curve. The unit is $(1/n_2 \times 10^{-3})$

We find that the linear phase δ_1 and δ_2 play an important role in output characters . Fig.(3.2) is plotted for $\delta_1 = \delta_2 = \pi/2$ and $5/8\pi$, the transmission versus nonlinear phase ϕ_2. In normal two-mirror FPI, altering linear phase only inflence transmission peak position,but for coupled nonlinear FPI, when linear phase altered , not only the peak

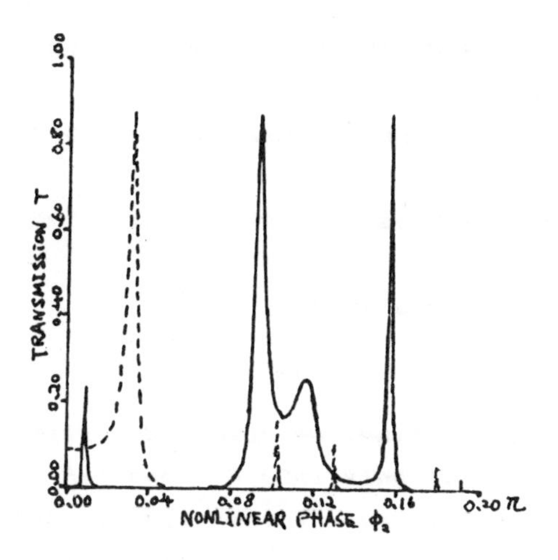

Fig (3.2) The transmission vs nonlinear phase ϕ_2 curve.

The real line for $\delta_1 = \delta_2 = \pi/2$, the dash line for $\delta_1 = \delta_2 = 5/8\,\pi$.

position, but also transmission peak number and profiles change. This give rise to the change of the input-output characteristics.

IV. Conclusion

In this paper, the coupled nonlinear FPI transmission formula has been deduced, the optical power input-output characters have been analysed using graphic method . It is found that linear phase δ_1 and δ_2 have great inflence on output characters. When δ_1 and δ_2 are proper slected, it is easier to realize multistability operation than normal two-mirror FPI, which can not realize multistability unless the change of nonlinear phase is greater than 2π.

REFERENCES :

[1] A.Sžoke et al. Appl. Phys. Letts. 376 , (1969)

[2] D.A.B.Miller, IEEE. JQE QE-17, 306, (1981)

NEGATIVE SLOPE BRANCH IN OB OF NONLINEAR MATERIALS

Chen Hong-pei, Zhang He-yi, Lian Gui-jun & Ai Bin

Dept. of Physics, Peking Univ.

Aug. 1987

Abstract *For the first time, we observed a stable negative slope branch in the OB of bulk ZnSe F-P and F-P filled with dye. In this paper, we'll present the experimental results and suggest that this phenomenon is caused by the sample's feedback change to the laser cavity.*

1. Introduction

In the process of our experiments on OB in bulk ZnSe F-P with uncoated surface, a new kind of phenomenon caught our interests, and we have also observed the same phenomenon in the F-P filled with HITC dye. The phenomenon presented in this paper, to our knowledge, was observed for the first time.

2. Brief Theory On the Instability of the Negtive Slope Branch In OB

As well known, the the negative slope branch shaped as 'S' in transmissive OB is thought physically unstable, no matter what the media is.

Felber and Marburger are apparently the first to use a graphical solution to obtain a state equation relating I_t and I_i. Suppose

$$\mathcal{T}(\beta) = \frac{I_t}{I_i} = \frac{1}{1+F\cdot\sin^2\beta/2} \qquad \text{..........(1A)}$$

and that the round-trip phase is given by $\beta = \beta_0 + \beta_2 I_i$.

The transmission also can be written as

$$\mathcal{T}(\beta) = \frac{I_t}{I_i} = \frac{\beta - \beta_0}{\beta_2 I_i} \qquad \text{..........(1B)}$$

For bistability to occur, there must be a region of three intersections of the Airy-function (1A) and the family of straight lines. (1B) (A→C region in Fig.1) This means that the slope of the curve exceeds that of straight lines.

$$\frac{\mathcal{T}(\beta)}{\beta} < \frac{d\mathcal{T}(\beta)}{d\beta} \qquad \text{..........(2)}$$

For a single characteristic response time system, According to Goldstone(Ref.2), we have

$$\tau \frac{d\beta}{dt} = \beta_2 I_i \mathcal{T}(\beta) - \beta + \beta_0 \qquad \text{..........(3A)}$$

and its differential form

$$\tau \frac{d\delta\beta}{dt} + [1 - \beta_2 I_i \mathcal{T}'(\beta)]\,\delta\beta = 0 \qquad \text{.........(3B)}$$

Pay attention to (1B), we can rewrite (2) as

$$1 - \beta_2 I_i \mathcal{T}'(\beta) < 0 \qquad \text{.........(4)}$$

For negative slope regions (4) applies, and (3B) have exponentially growing solutions.

Hence, the negative slope regions are unstable. By identical arguments, the positive slope portions are stable.

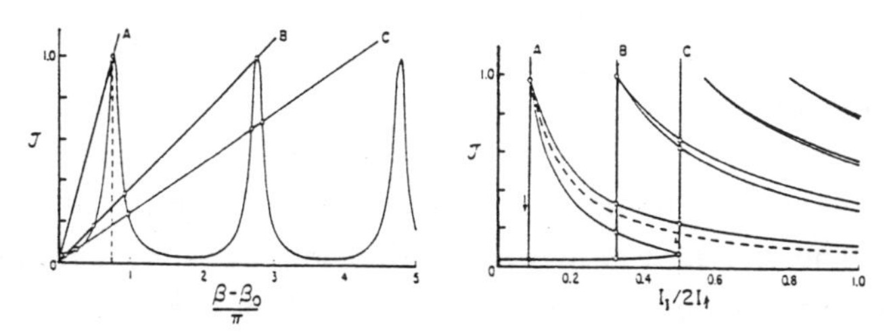

Fig.1. Graphical solution showing the occurance of the optical bistability.

In our experiments, however, in both bulk ZnSe F-P and in F-P filled with dye, we have observed stable negative slope $dI_t/dI_i < 0$ branches in I_t verses I_i curves and in the corresponding branch of I_r verses I_i curves(Fig.3 & 4).

3. Experimental Setup and Results

Fig.2 shows our experimental setup.

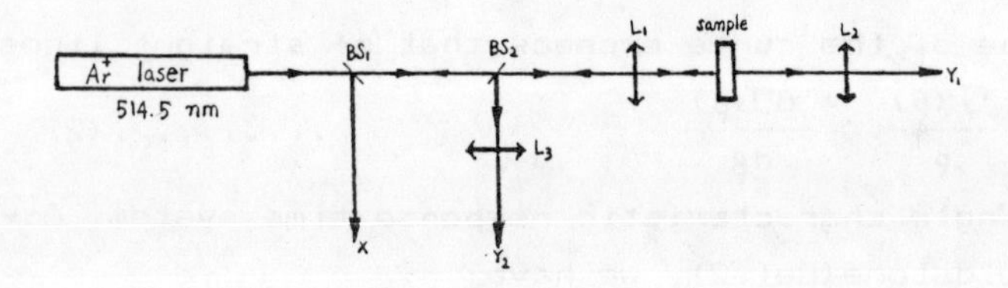

Fig.2 Experimental Setup

Our ZnSe sample was 800 um thick, CVD-grown, and polished plane parallel to form a natural-reflective, low-finesse F-P etalon, with imparallelism less than 1 mrad. The etalon full of dye solution, monofilm reflectivity being 50%, had a thickness of about 670 um, and had a dye concentration of 5×10^{-5}-- 1×10^{-4} mol/l. Operating wavelength was 514.5 nm.

Fig.3&4 shows the $dI_t/dI_i<0$ negative slope branch in bulk ZnSe F-P and in the F-P full of dye respectively. The 'S' shaped branch in I_t-I_i curve indicated a decrease of incident power with a meanwhile increase of the transmission at the occurrance of OB.

Experiments also shown that the negative slope branch was stable instead of unstable. We held the incident power at the midway in the negative sloped portion and saw the system held the present state for as long time as expected. No abrupt state change occured. This fact proved that the system was stable in negative slope portion.

4. Formation of the 'S' Shape Branch

One can easily understand the 'S' shape branch in view of feedback. In Fig.2, the lens L_1 served as a Gauss beam

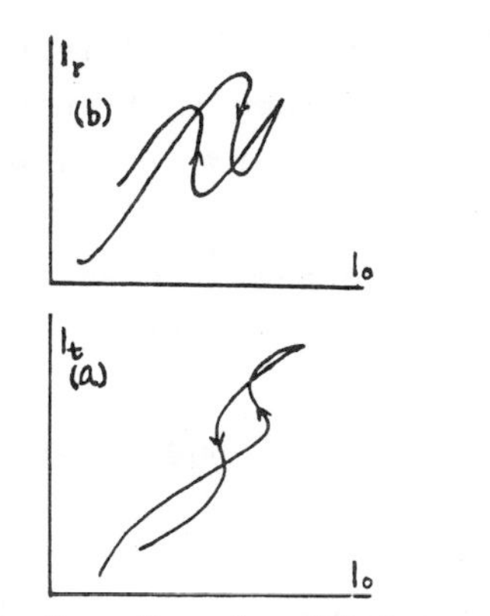

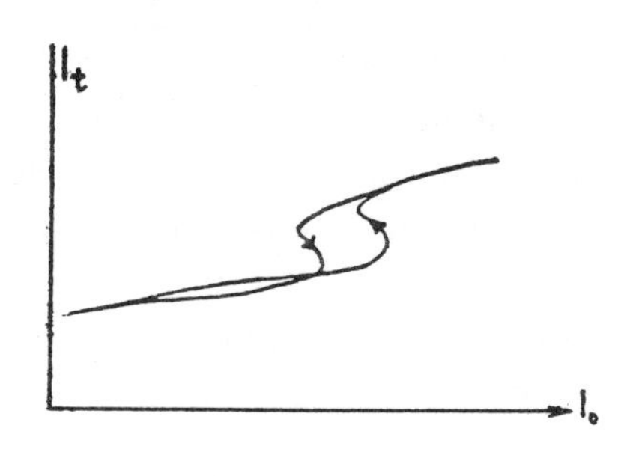

Fig.3 Experimental transmission(a) reflection(b) nonlinear response in bulk ZnSe

Fig.4 Experimental transmission nonlinear response in HITC dye solution

waist translocator. The beam waist was translocated to meet the front surface of the sample, or in dye solution case, the front reflective surface of the F-P etalon, and meanwhile a normal incidence was adapted. (We label the above condition as 0°condition). Because the wavefront of the Gauss beam waist was a plane one just to coincide with the front surface of the sample, the reflective (feedback) beam would exactly coincide with the incident one and give out a most effective feedback to the laser cavity.Otherwise the feedback would be much smaller than that in 0°condition.

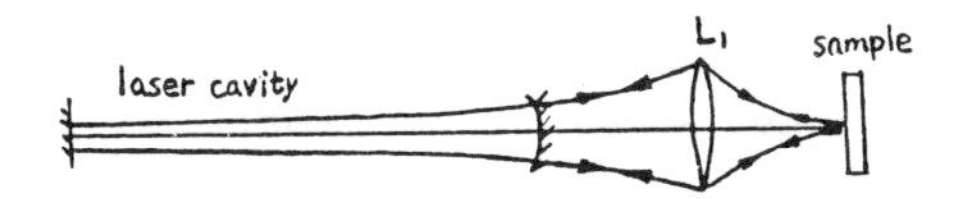

Fig.5 In 0°condition, feedback beam coincide with incident beam.

At the occurrance of OB switchings, with the turning-on of the transmission, there occured a reflective(feedback) turning-off(or decrease). And thus the laser output, i.e. the incident power of the sample, decreased, while the transmission of the F-P cavity still held high. Now that the 'S' shape shaped. On this point of view,we could expect that the higher the sample etalon's finesse, the clearer the 'S' shaped phenomenon would be. In fact, one can find in Fig.4 that the 'S' shape is much more obvious than that in Fig.3, because the filmcoated etalon had much greater finesse than the bulk ZnSe natural etalon.

We have the following evidences to show the action of feedback in the formation of 'S' shape branches:

5. Experimental Evidence of the Feedback Change Causing the Formation of 'S' Shape Branches

(1) Feedback at various incident angles

Fig.6 shows an angular distribution of the sample's feedback, and a maximum at 0° incident angle (0° condition). So certain incident angles correspond to certain feedback.

(2) Influence of the feedback to laser output

In Fig.7, the influence of the feedback to the laser output irradiance is shown obviously. The X-axis is the incident angle, and y-axis is the laser output. The parameter is the plasma tube exciting current. In receiving of the laser output with feedback, we have observed an about several hundred Hz unstable signal which changed irregularly with the increase and decrease of the incident power.

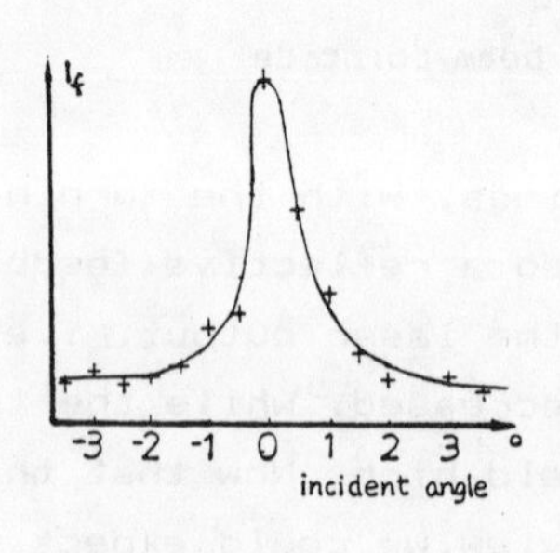

Fig.6 Angular distribution of the feedback.

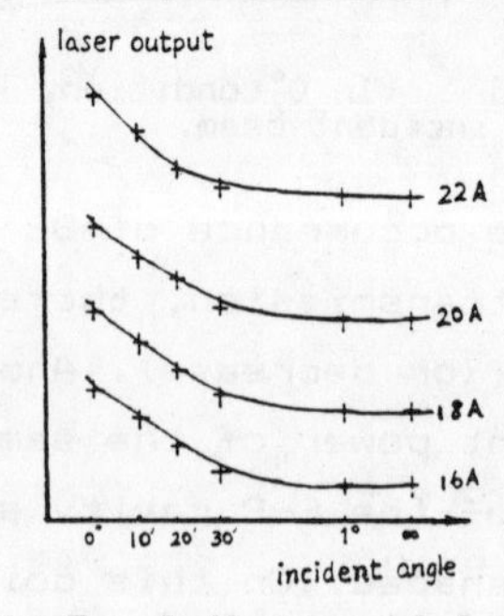

Fig.7 Influence of feedback to the laser output.

(3) Feedback influenced the negative slope branch

We can find in Fig.6 that the increase of incident angle meant the reduction of feedback. Fig.8 shows that the negative slope branch declined with the reduction of feedback.

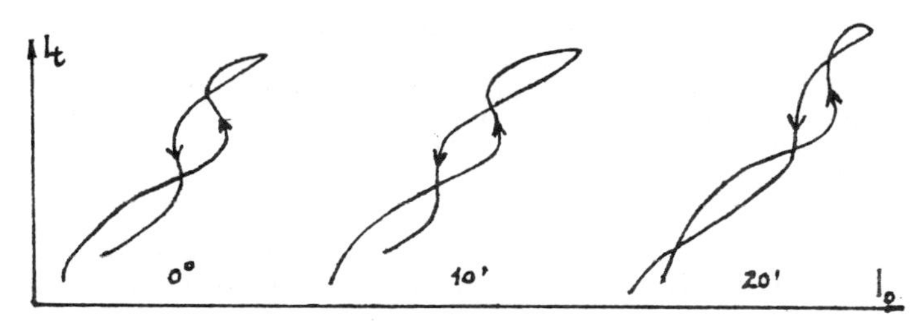

Fig.8 Recession of the negative sloped branch with the reduction of feedback.

6. Further discussion on the experiments

Although our experiments shown that the negative slope branch in transmissive OB exist and be stable, and that the turning-on and -off branches of the hysteresis can reach each other very near(Fig.9). We still cannot make sure that this do be the 'S' shape which Goldstone dealt with, for all the analysis of Goldstone are based on the suppose of single characteristic response time system. While our system has at least two incomparable characteristic response time, one is the sample's thermal response time, which is about 10^{-3} sec, and the other is the laser cavity's response time for building the feedback re-oscillations in the cavity, which is about 10^{-6} sec. We haven't got evidence to show that Goldstone's analysis can apply in two or more characteristic response time system. Maybe the existance of the sample's feedback changes the system's behaviour, and we should develop a further theory to demonstrate the stability of the 'S' shape branch in such systems theoretically. Further experimental and theoretical investigations are undergoing.

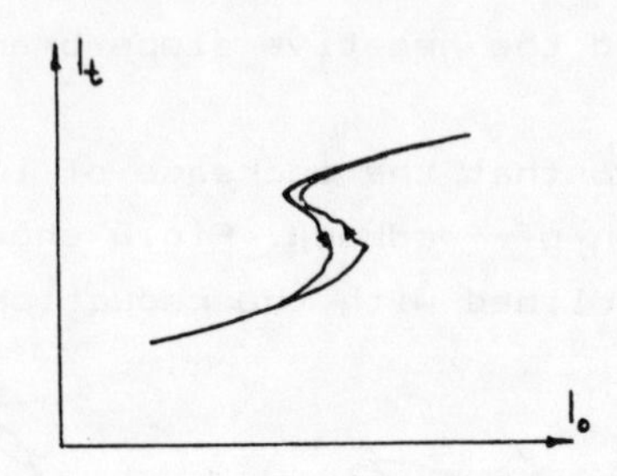

Fig.9 The turning-on and -off branches of the OB hysteresis reach very near but do not coincide with each other. Experiments in alcohol solution of HITC dye.

Last we would say that the 'S' shape branch stability refer to the stability of the whole system instead of the sample's OB alone. Because of the feedback,we can't apart the sample from the system. The present paper suggests a way to obtain a negative 'optical impedance' character, and it would be of some use for optical computing or for some hybrid OB system. Further theoretical and experimental works concerned are undergoing.

Acknowledgement

We would like to express an appreciation to Beijing Institute of Artificial Crystal for their great helps in sample providing.

References

1. Felber and Marburger. Appl.Phys.Lett. 28,731 (1976)
2. H.M.Gibbs. OB. Academic Press,1985, Section 6.1--6.4
3. Goldstone Laser Handbook, Vol.4
4. B.S.Wherrett J.O.S.A. 345--350 Vol.3 No.2 Feb. / 1986
5. D.A.Miller. IEEE.J.QE,Vol.QE-17,306,No.3 Mar./1981
6. Wherrett et.al Optics Commun. 52,301,1984
7. Wherrett et.al J.O.S.A. 351,Vol.3, No.2 Feb./1986

Optical Bistability in Very Thin ZnSe Film

Sun Yinguan, Jiang Ziping, Xiong Jun, Wang Weiming

Physics Department, Beijing Normal University

Guo Ping, Shang Shixuan

Analytical and Testing Center, Beijing Normal University

We report the demonstration of optical bistability (OB) and transverse effects in very thin ZnSe film . The results are consistent with the observed thermal shift of interband absorption.

Optical bistability due to increasing absorption was first discussed by Kaplan[1] and Hajto et al[2].Such OB has also been demonstrated in a number of materials, for instance , amorphous GeSe film [2] , CdS,CdSe [3] slice, GaAs/AlGaAs superlattice [4] , InSb [5] and ZnSe[6]. Although light induced absorptive OB has been observed in bulk ZnSe by Taghigadeh et al[6], we have made ZnSe film samples by vacuum deposition with thermal source and investigated some very thin (10μm, 6.5μm, 4.4μm and 0.4μm) ZnSe film samples at several output wavelengths of Argon-ion laser. The samples show strong nonlinear phenomena and demonstrate OB in mirrorless system.

The experimental setup is shown in Fig.1. The beam from an Ar+ laser is focused on the sample within a spot of 60 μm in diamter. The laser intensity is adjusted by a variable neutral density filter . The incident and transmitted intensity are detected by photodetectors . The experimental results are shown in Fig.2 and Fig.3 . It can

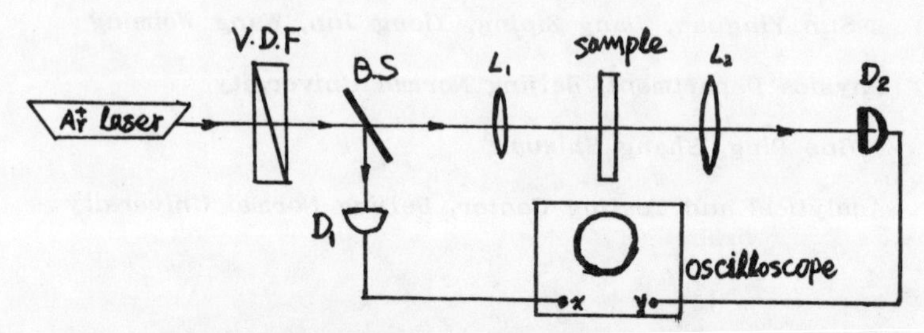

Fig.1 Experimental setup . BS--beamsplitter ; L--lens ; D--photodetector ; VDF--variable density filter.

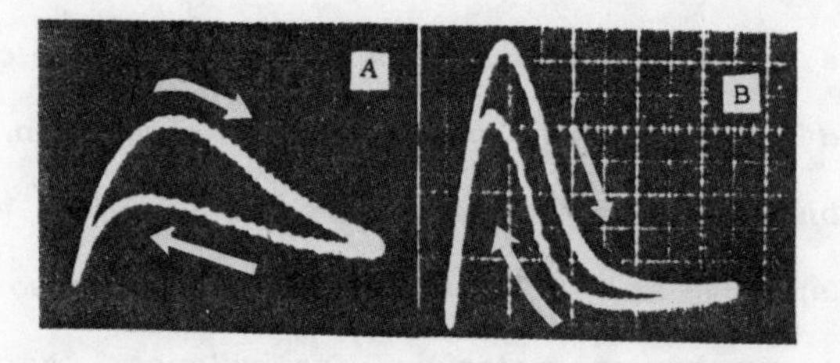

Fig.2 The optical bistablity of ZnSe film (4.4 μm) with laser wavelength of 4880 Å . The incident intensity is 20mw/div . (A)dI/dt=200mw/sec; (B)dI/dt=50mw/sec

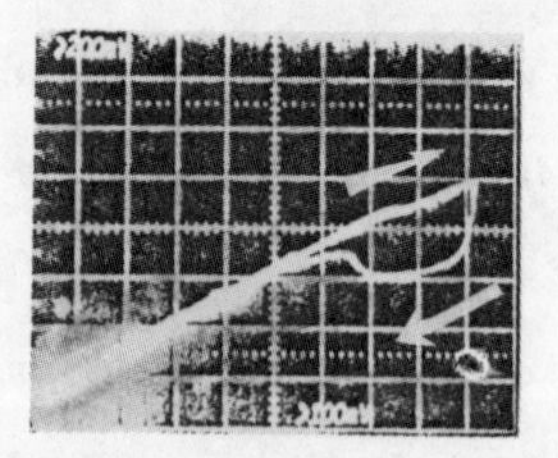

Fig.3 OB of ZnSe film (0.4 μm) with laser wavelength of 4765 Å. The incident intensity is 50 mw/div, dI/dt = 500 mw/sec

be seen that the laser beam heated the sample to increase its temperature which in turn leads to the increasing of the absorptive coefficient . This results in a more increasing heating. So when decreasing the light intensity , the absorption is still high for the high temperature of the sample and the transmitted light intensity is on the lower branch of the hysteresis. Our observations show that the shape of the hysteresis is strongly related to the rate of the modulation of the input power and not sensitive to the thickness of the sample, the wavelength and the angle of the incident light. The measurements of the samples absorption at different temperatures, as shown in Figure 4 , prove our analysis .

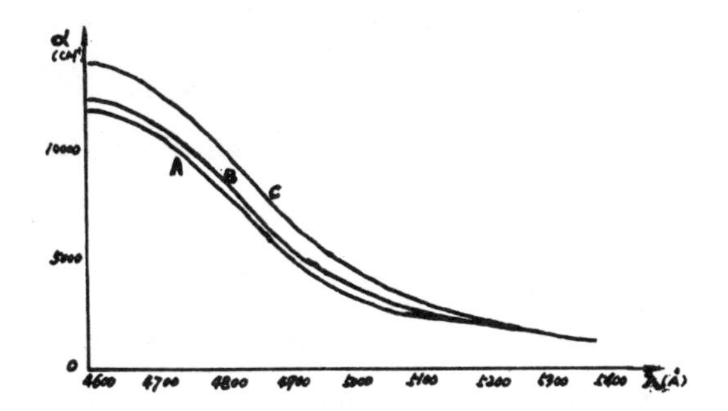

Fig.4 The absorption curves of ZnSe film at different temperatures. (A). T=2 °C (B). T=43 °C (C). T=85 °C

Because of the Gauss beam of the input light,there would be different temperature distribution within the spot on sample and this results the transverse effects of the transmitted light as shown in figure 5 at different incident power . Our observation suggest that

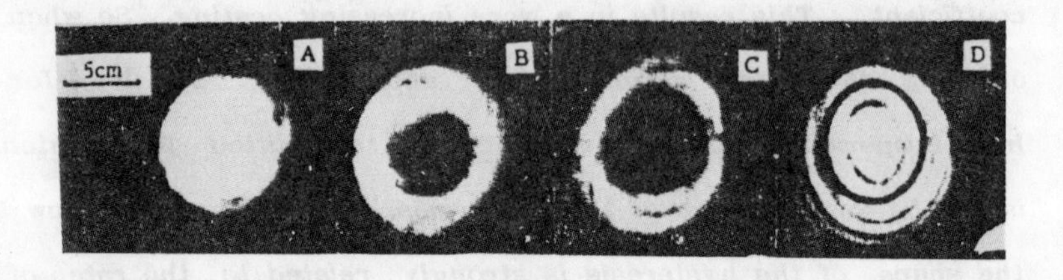

Fig.5 The cross sections of the transmitted light in far-field (one meter away from the 4.4μm sample) at different incident light intensity. The beam of the laser (wavelength of 4880A) was focused onto the 4.4um ZnSe sample with a f=5cm lens and at normal intensity . (A) The laser spot ; (B) I = 50 mw ; (C) I = 100 mw ; (D) The situation after the sample damaged.

before laser induced damage takes place probably due to the decomposition of ZnSe at high input power , ZnSe thin film would be a stable OB plate under milliwatt power level , and thin ZnSe film would also be an excellent sample for the study of nonlinearity.

Wherrett et al[6] showed that,for an exponential band edge which shifted linearly with increasing temperature , bistability should be observable under the condition:

$$aL<0.18$$

where a is the initial absorptive coefficient of the sample of thickness L and has been experimentally verified for the case of InSb. In our experiment , however, aL=0.2--8, this is clearly contradicting with the model . We think it is impossible to explain this results by using self-focusing as in bulk ZnSe material[6]. It must consider the complex structure of the thermal deposition material.

We gratefully appreciate the useful disscussion with prof. Zhang Heyi, Physics Department of Peking University.

REFERENCES:

1. *A.E.Kaplan: Phys.Rev.Lett.48,138(1982)*
2. *J.Hajto, I.Janossy: Philos.Mag.B47,346(1983)*
3. *M.Dagenais ,W.F.Sharfin: Appl.Phys.Lett.45,210(1984)*
   P.Lavallarcl, P.H.Duong, T.Itoh: Phsica B 117&118, 410(1983)
4. *D.A.B.Miller, A.C.Gossard, W.Weigmann: Opt.Lett.9,102 (1984)*

5 *B.S.Wherrett, F.A.P.Tooly, S.D.Smith: Opt.Comm.52,4,301(1984)*

6 *M.R.Taghizadeh,I.Janssoy,S.D.Smith: Appl.Phys.Lett.4,46,331(1985)*

7 *Jiang Ziping : M.S.Thesis (1986), Beijing Normal University*

OBSERVATION OF OPTICAL BISTABILITY OF ELECTRONIC ORIGIN IN CdHgTe

Zhang Jianguo, Chen Jishu, Zheng Kangli, Hu Quanyi,

Optics Department, Shandong University, Jinan, China.

North China Research Institute Of Electro-Optics, Beijing, China.

I. INTRODUCTION

The alloy semiconductor $Cd_xHg_{1-x}Te$ offers a variation of band gap with composition from 0 to 1.5 eV. Low-temperature CdHgTe shows an extremely large optical nonlinearity of electronic origin at 10.6 μm for samples with band-gap energies close to this photon energy[1]. However, it seems difficult to achieve optical bistability utilizing this kind nonlinearity with a CW CO_2 laser because of the Auger recombination and the thermal effects of CdHgTe[2]. Here we report our preliminary observations on optical bistability of electronic origin by using a fast-modulated quasi-CW laser light.

II. EXPERIMENTAL DETAILS

A sample of p-$Cd_{0.221}Hg_{0.779}Te$ with a carrier concentration 1.9×10^{16} cm^{-3} was polished to form a plane parallel etalon of thickness 400 μm and was held in a cryostat at 77 K. The band gap resonant wavelength is 9.67 μm. Radiation from a 8 W line tunable CO_2 laser operating in the 10.22 μm wavelength band after passed through an attenuator, a modulator with modulation frequency 1100 Hz and a beamsplitter was focused onto the sample by a Ge lens. The spot diameter was about 100 μm (1/e points). The intensities of the incident and transmitted beams were monitored with two small area, fast response, 77 K CdHgTe detectors positioned at the centers of the beams. The incident and transmitted signals were observed on an oscilloscope.

III. RESULTS AND DISCUSSION

Fig. 1 shows the shape of incident and transmitted pulses. The risetime of incident pulse is 100 μs. The transmitted pulse shows switching behavior and the switch-on time is about 10 μs. The measured incident peak power was about 300 mW.

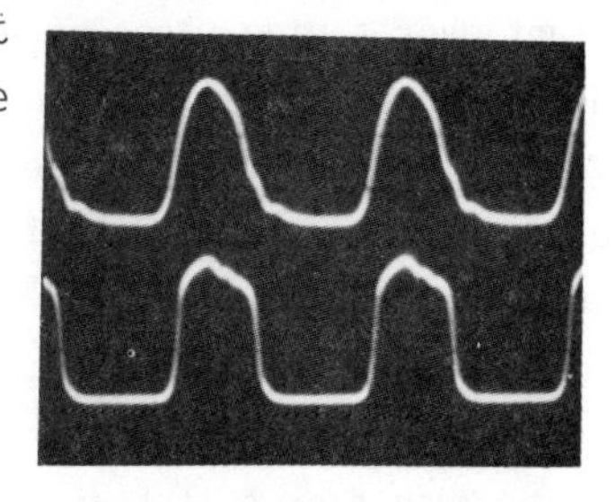

Fig.1 Incident power(20 mV-/div) and transmitted power (20 mV/div) against time (100 μs/div).

Fig. 2 shows the input-output characteristic. Optically bistable action with the hysteresis loop clearly appears. The switch-on intensity is about 1.5 kW/cm^2.

The electronically induced switching in optical bistability is dependent on irradiance and not on energy flux[3]. We have changed the modulation frequency from 500 Hz to 1100 Hz and the space ratio of the modulator from 3:1 to 6:1 and 12:1 so that we changed the energy flux, no changes in the input-output characteristic due to thermal effects were observed.

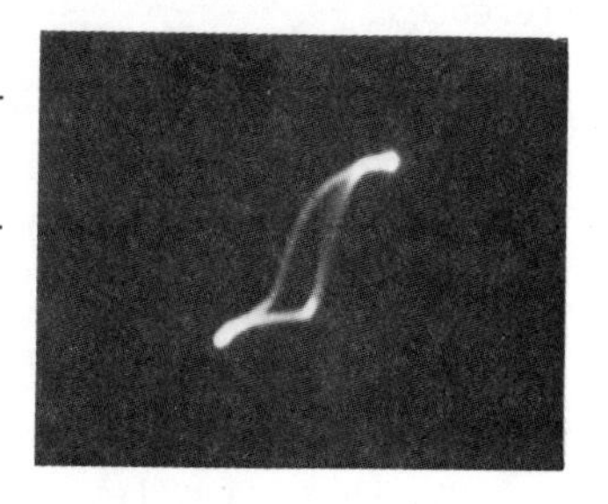

Fig.2 Input-output characteristic showing optical bistability.

Interband absorption of the laser beam generates free electrons, which rapidly thermalize to a quasi-equilibrium near the band edge and modulate the interband dielectric function, resulting in optical band-gap resonant nonlinearity[4]. In n-type or intrinsic CdHgTe, the interband recombination time t_R is Auger limited and decreases rapidly as the excess carrier density increases at high pump level[5], so that the optical nonlinearity rapidly limited[2] at intensities above about 10 W/cm^2.

This fact, combined with the thermal effects, means that optical bistability of electronic origin is difficult to achieve using a CW laser[2]. This model does not apply to the present experiment with p-CdHgTe because the interband recombination in p-CdHgTe is dominated by Shockey-Read trap centers and is constant over a wide range of excess carrier densities[6] and, hence, intensities. The Shockley-Read recombination time is inversely proportional to the acceptor concentration[6]. The band-gap resonant nonlinearity in this material is termed the dynamic Burnstein-Moss shift or the band-filling effect[7].

As shown above, the thermal effects were avoided because of using a fast modulated laser light in our experimental arrangement. Much fast switch-on time should be possible if more intense and fast-rising incident pulse could be available, as accomplished in InSb[8].

References:

<1> Jain, R.K. and Steel, D.G., Opt. Commun. 43, 72(1982).

<2> Miller, A., Parry, G. and Daley, R., IEEE J. Quantum Electron. 20, 710(1984).

<3> Tooley, F.A.P., Walker, A.C. and Smith, S.D., IEEE J. Quantum Electron. 21, 1340(1985).

<4> Miller, D.A.B., Seaton, C.T., Prise, M.E. and Smith, S.D., Phys. Rev. Lett. 47, 197(1981).

<5> Bartoli, F., Allen, R., Esterowitz, L. and Kruer, M., J. Appl. Phys. 45, 2150(1974).

<6> Polla, D.L., Tobin, S.P., Reine, M.B. and Sood, A.K., J. Appl. Phys. 52, 5182(1981).

<7> Yuen,S.Y. and Becla, P., Opt. Lett. 8, 356(1983).

<8> Walker, A.C., Tooley, F.A.P. and Smith, S.D., Phil. Trans. R. Soc. Lond. A313, 249(1984).

OPTICAL BISTABILITY AND OPTICAL LOGIC BY THE NEGATIVE RESISTANCE EFFECT IN AN AVALANCHE HETEROJUNCTION PHOTOTRANSISTOR

HUANG XIAO-KANG, DONG JIE
DEPT. OF INFORMATION ELECTRONICS
TSINGHUA UNIVERSITY
BEIJING, P. R. CHINA

Abstract: The optical bistability and some types of optical logic gates, such as OR,NOR,AND,NAND and INVERTER were obtained due to the negative resistance effect of an avalanche heterojunction phototransistor is reported for the first time.Since having very simple structure and to be very easy to integrate, this device could be used in digital optical signal processing.

In this paper, we demonstrate a new kind of optical bistable scheme based on the avalanche heterojunction phototransistor (AHPT) and the semiconductor laser diode.Using this scheme, we have realized some types of optical logic gates, such as OR,NOR, AND,NAND and INVERTER.

The AHPT was grown by the LPE technique in our lab with the following five layers:(1) Buffer,n^+-GaAs:Te($2*10^{18}cm^{-3}$);(2) Collector,n-GaAs:Sn($2*10^{17}cm^{-3}$);(3) Base,P-GaAs:Ge($0.7-1.8*10^{18} cm^{-3}$); (4) Emitter,n-$Ga_{0.55}Al_{0.45}As$:Sn($3*10^{17}cm^{-3}$); (5) Cap,n^+-GaAs:Te($2*10^{18}cm^{-3}$).In order to get the negative resistance, it is very important to adjust the carrier concentration in the different layers, especially in the base and the collector. In our work, the width of the negative resistance region varies from 0.4V to 4.0V. Fig.1(a) shows the I-V characteristic of the device. V_p (in Fig.1(b)) is the peak voltage which is the critical point between the normal region and the negative resistance region. V_p decreases with the increase of the incident light power P_{in}, i.e. the I-V characteristic moves towards left as the incident light power P_{in} increases.

In order to show the bistable property of the phototransistor, connect it with a load in series to form a nonlinear system, for example, a resistor or a semiconductor laswe diode. AB is the load line, which has intersections with the I-V curves of the phototransistor. Different intersections represent different states. Point 1,2,and so on are the stable state at the OFF state. As the incident light power increases to a certain value, the current through the device becomes very large, i,e. it jumps from the point 3 to the point 4. Now, with the reducing of the incident power, V_p will increase and the I-V curve will move towards right. The intersection will go back along AB to the point 5 and 6. We define the point4,5 ... as the ON state and the point 1,2 ... as the OFF state. Fig.2 shows the bistable hysteresis loop of the bistable system. When the device works at the ON state, the output light power is about 1 mW and the ratio of the output power at the ON state to that at the OFF state is larger than 4dB.

Due to the negative resistance effect, we have experimentally realized several kinds of optical logic gates and switching circuit. With the experimental set-up as shown in Fig.3, and putting into two independent optical signals, when the two signals satisfy a certain relation with the critical input power, OR gate and AND gate can be obtained, respectively.Fig.4 illustrates the INVERTER circuit. The output signal is out of phase with the input signal. Similarly, NOR gate and NAND gate can also be obtained. Fig.5 shows the logic relation between the input and the output signal of OR gate and NAND gate.

The response time of the bistable system is limited by two factors. One is the response speeds of the laswe diode and the phototransistor, the other is due to the critical slowing down. While jumping from the OFF state to the ON state, and jumping from the ON state to the OFF state, the delay time will become very long. We observed the delay time to be in the range from tens of microseconds to tens of millseconds. Fig.6 shows the experiment result of both the up and down critical slowing down observed simultaneously. In order to reduce the delay time,the input light power should be as large as possible.

The advantage of this device over the conventional ones is that no external feedback is needed. Therefore, it is possible and convenient to realized either monolithic or hybrid integration, and it is promising for using in practical systems.

Figure Captions:

Fig.1(a) The I-V characteristics of the AHPT under different input optical powers.With the increase of P_{in}, the curve moves towards left.(I:0.5mA/div;V:0.5V/div)

(b) The schematic showing of the different intersections representing different states.

Fig.2 The bistable hysteresis loop of the bistable system.

Fig.3 The experimental set-up for OR gate and AND gate.

Fig.4 The experimental set-up for the INVERTER.

Fig.5 The logic relation between the input and output signals of (a) OR gate; (b) NAND gate.

Fig.6 The critical slowing down of both the up and down of the bistable system observed simultaneously under different input levels P_{in1} P_{in2} P_{in3}.

a

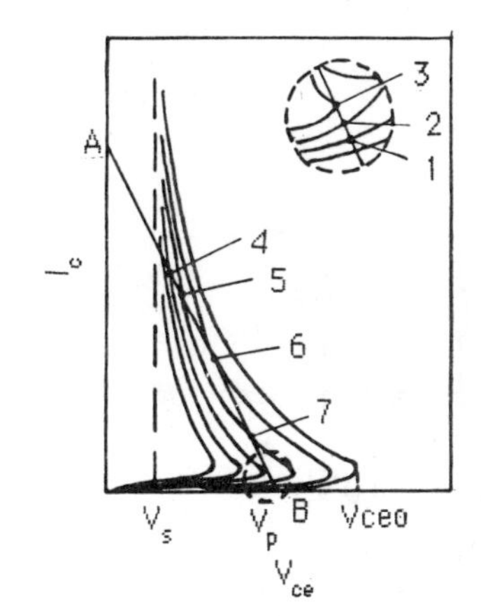

b

Fig.1

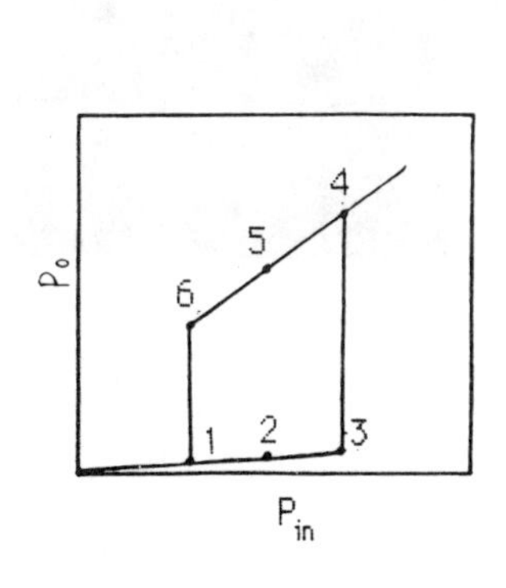

Fig.2

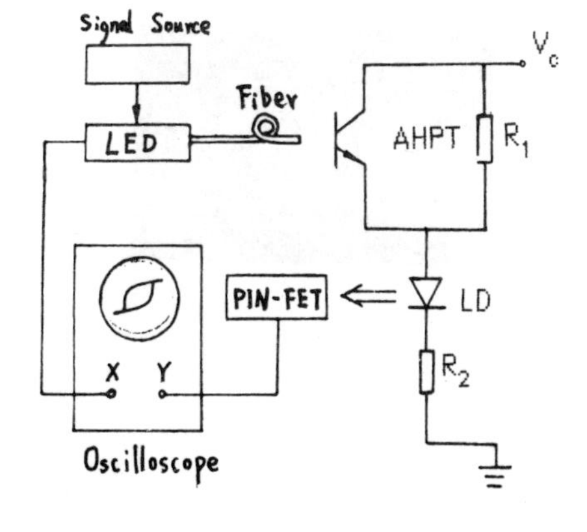

Fig.3

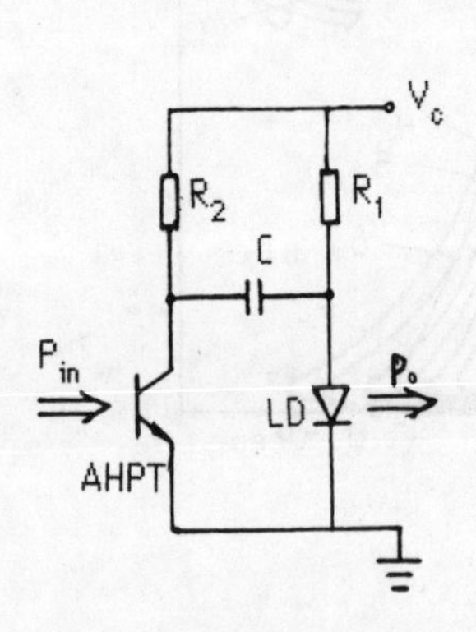

Fig.4

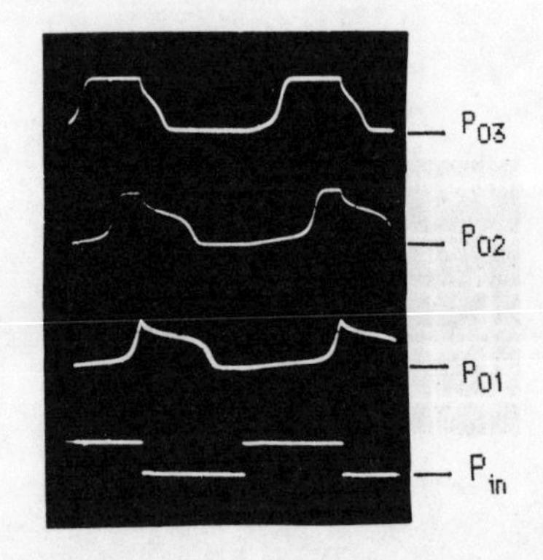

Fig.6

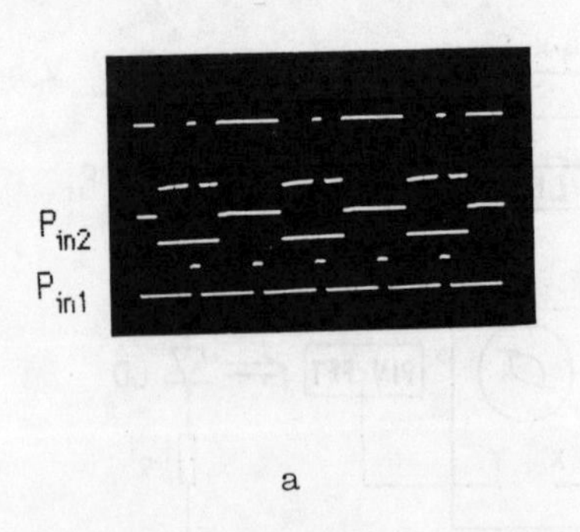

a

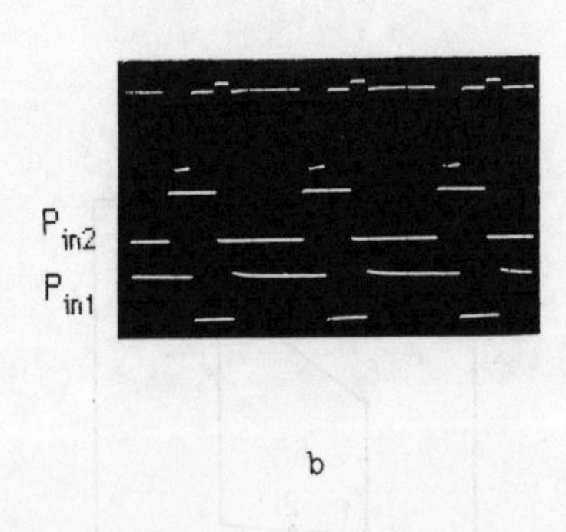

b

Fig.5

BISTABILITY IN SEMICONDUCTOR LASERS WITH INHOMOGENEOUS CURRENT INJECTION

Niu Jin-zhen
Dept.of Physics, Central Institute for National Minorities,Beijing, China
Chen Wei-xi
Dept. of Physics, Peking Univ.Beijing,China

Abstract
In this report we discuss bistability of CW-AlGaAs BH laser with a segmented contact. The dependence of bistablity and switching on the photovoltaic effect is calculated.

There has been considerable interest in bistable optical devices which are to perform optically controlled memory and switching operation,and which may be crucial to the success of all-optical systems. The ordinary method of achieving optical bistability involves combining a medium ,which displays intrinsic nonlinear reflection with a Fabry-Perot cavity[1].On the other hand, the use of inhomogeneously excitied semiconductor lasers for achieving optical bistability has also been to be a uniquely noteworthy method[2],[3].they denmonstrate successfully that this bistable laser can be used as an optical stylus for optical disk readout with an excellent signal to noise ratio and a large electrical output signal.

In this report we discuss bistability in CW-AlGaAs buied heterostructure(BH) lasers with a split contact on the top p-side as shown in Fig.1.The lasers have been fabricated by a standard process, typical dimensoins and doping levels are given in the figure. The upper cladding layer of These BH lasers is only lightly p-doped in order to increase the lateral parasitic resistance between the two contact pads.

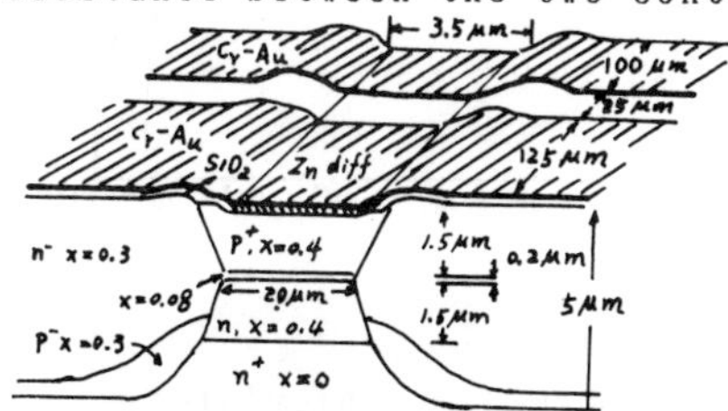

Fig.1.$Ga_{1-x}Al_xAs$ BH laser with a segmented contact

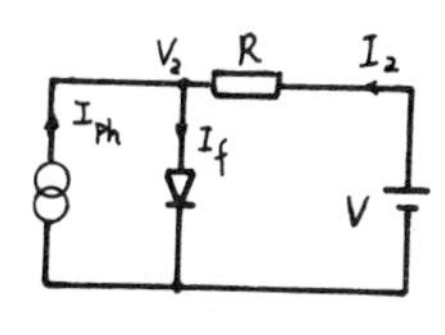

Fig.2.Equivalent circuit of the photodiode

The light-current characteristic is calculated using familiar rate equations:

$$dN_1/dt=I_1/(qc_1v)-R(N_1)-G(N_1)S \quad (1)$$

$$dN_2/dt=I_2/(qc_2v)-R(N_2)-G(N_2)S \quad (2)$$

$$dS/dt=[c_1G(N_1)+c_2G(N_2)-1/\tau_p]S+m[c_1R(N_1)+c_2R(N_2)] \quad (3)$$

These are for the electron density in the gain section,the

electron desity in absorser section and the photon density in the lasing mode, respectively. $v_1 = c_1 v$ is the volume of the gain region, N_1 is the inversion density of carriers in the gain region, and I_1 is the current injected into it. $c_2 v$, N_2, and I_2 are correspon-ding variales in the absorber scction; q is electronic charge, τ_p is the photon lifetime and m is the coupling coeffcient for spontaneous emission into the lasing mode. S is the photon density. The spontaneous emission rate is taken to be $R(N)=B(N+N_A)N$ where B is recombination costant and N_A is the doping concentration of the p-type background in the active region. A linear gain dependence on injected carrier density $G(N)=A(N-N_{tr})$ is used in the following calculation. In the formulations single-transverse mode and single longitudinal mode oscillotion are assumed.

In some calculations the current into the absorber section is assumed a constant. In fact, I_2 is dependent on the density of photons in the active region.

Consider the bistable laser consisting of two parts, a gain section which is pumped with a constant current I_1 and an absorber section which acts like a photodiode with the optical cavity. The current I_2 through this photodiode consists of two terms: the first term is the normal diode current which depends exponentially on the applied voltige V_2 while the second term correspounds to a negative photo-induced current I_{ph} which is proportional to the photon density in the active region (as shown in Fig.2).

The dependence of bistability and swicthing on the photovoltaic effct is discussed in the report.

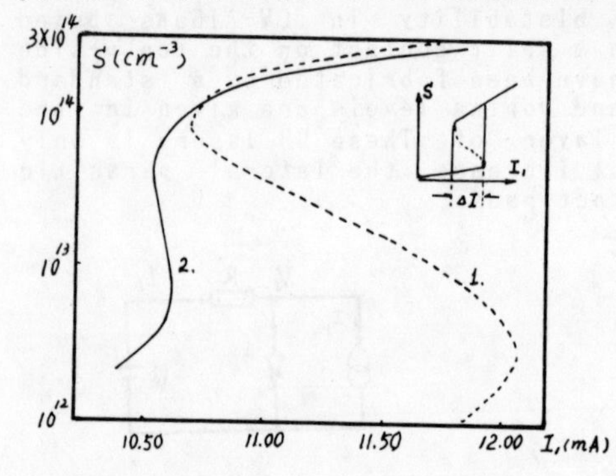

Fig.3. Calculated static S/I_1 characteristic for two different bias conditions of absorber section. 1. R=46kΩ, V=5v, 2. R=100Ω, V=1.42v.

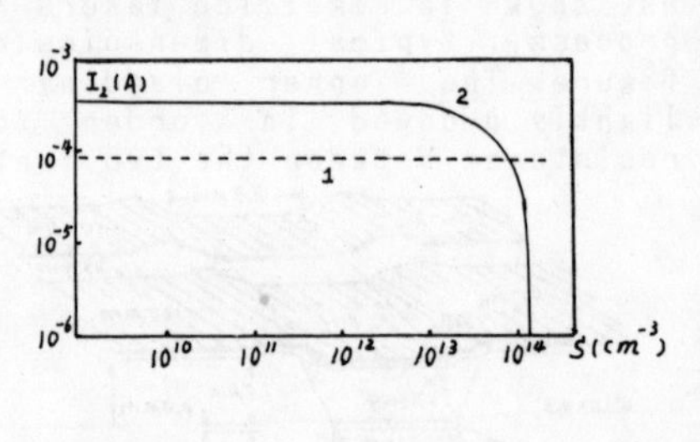

Fig.4. Calculated I_2/S characteristic for two different bias condition of absorber section. 1. R= 46kΩ, V=5v, 2. R=100Ω, V=1.42v

1. Static Analysis

The light-current characteristic is given in Fig.3 for two different bias conditions of the absorber section from (1)-(3). The current I_2 is calculated by the following formulas:

$$V_2=V_g+V_T\{Ln[N_2(N_2+N_A)/(N_c N_V)]+0.25\sqrt{2}N_2(1/N_c+1/N_V)+0.25\sqrt{2}N_A/N_V\} \quad (4)$$
$$I_2=(V-V_2)/R \quad (5)$$

where R, V, are the resistance and voltage of the bias circuit of the absorber section(as shown in Fig.2). The external voltige V_2 across the absorber section is given using an approximation for the carrier density in a parabotic band. The following values of the various parameters have been used: $c_1=c_2=0.5$; $v=250\times2\times0.2\mu m^3$; $B=3\times10^{10}cm^3/s$; $N_A=3\times10^{17}cm^{-3}$; $A=2.34\times10^{-6}cm^3/s$; $N_{tr}=0.83\times10^{18}cm^{-3}$; $\tau_p=1.5ps$; $m=10^{-4}$; $V_T=25.9mv$; $V_g=1.46v$; $N_c=4.7\times10^{17}cm^{-3}$ and $N_v=7\times10^{18}cm^{-3}$. The dashed line corresponds to biasing the absorber section with a large resistance(R=46kΩ, V=5v), A remarkable hysteresis loop is seen in the curve.the solid line corresponds to biasing the absorber section with a small resistance(R=100Ω;V=1.42v), only has a very narrow hysteresis.

Fig.4 calculated I_2/S characteristic for two different bias conditions of absorber section. The dashed line corresponds to biasing the absorber section with a large resistance(R=46kΩ, V=5v) the I_2 is stable with incseasing of photon density.The solid line corresponds to biasing the absorber section with a small resistance(R=100Ω, V=1.42v).It can be seen that the I_2 decreases rapidly with the increasing of photon in the heigher photon density.

Obviously, The bistability current range(ΔI in the inset) corresponds to the bias condition of the absorber section.Perhaps, there are additional electrical coupling which consists of the following two parts. A constant parasitic resistance is due to the finite lateral conductance of the p-type top cladding layer and a connection through carreirs can be generated in the optical waveguide connecting gain and absorber section which looks like photoconductive effect. So When the bistable laser is switched off,the resistance of this photoconductive path is very large. But if it is switched on the carrier density within this section is increased to transparancy level and the resistance is dropped.

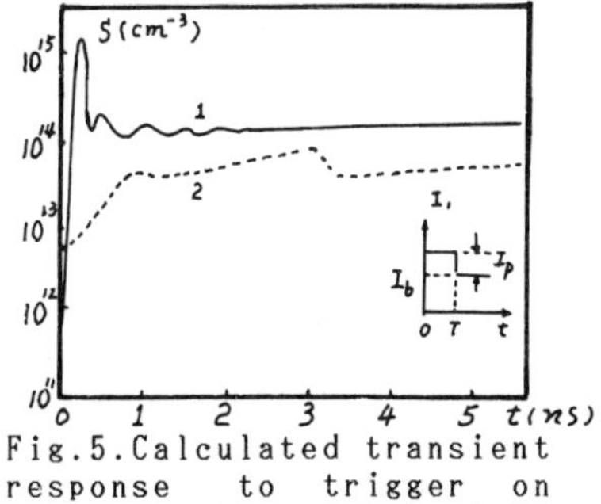

Fig.5.Calculated transient response to trigger on current pulse with $I_p=5\Delta I$,
1. R=46kΩ, V=5v, T=0.5ns,
2. R=100Ω, V=1.42v, T=3ns.

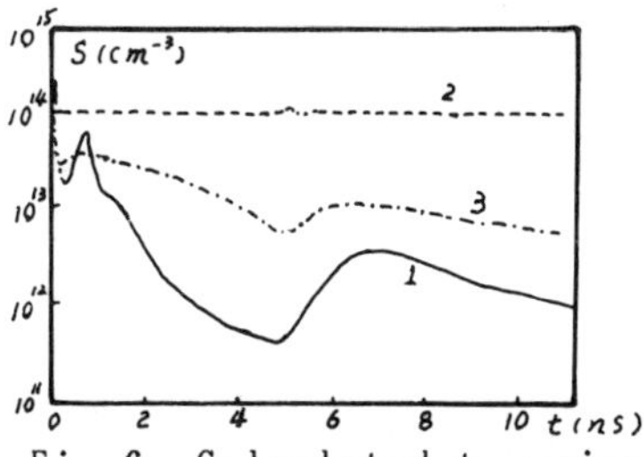

Fig.6. Calculated transient response to trigger off current pulse with T=5ns.
1.R=46kΩ, V=5v, $I_p=-1.4\Delta I$,
2.R=100Ω, V=1.42, $I_p=-1.4\Delta I$,
3.R=100Ω, V=1.42v, $I_p=-5\Delta I$.

2.Switching

the influnence of the photonvoltaic effect for switching from an off state to an on state and from on to off state has also been calculated. Photon response to 5ΔI height trigger current on pulses, with different widths, are shown in Fig.5 for different bias condition of the absorber section. For large resistance, with 0.5ns current pulse, switching can be carried out, but for small R, the width of pulse which can switch from off to on is greater than 3ns.

Responses to off pulse are shown in Fig.6. For large R when I_p=-1.4ΔI a 5ns-wide current is enough for switching, but for small R switching can not be carried. If the height of pulse is increased to I_p=-5ΔI, switching from on to off can be achieved.

From these calculations it can be seen that the photovoltaic effect affects the switching process of the bistable laser. When the resistance of bias circuit of the absorber section or the equivalent resistance between the gain section and the absorber section is small the influence of the photovoltaic effect is obvious. Th experimental results will be reported later.

REFERENCES

[1].H.H.Gibbs, Opt. News Summer, 6(1979)

[2].K.Y.Lau, A. Yariv, "Bistability and Pulsation in Semiconductor Lasers with Inhomogeneous Current Injection" IEEE J. Quantum Electron. vol. QE-18, pp.1351-1360, Sept. 1982.

[3].H.Kawaguchi, "Optical bistabe-switching operation in semiconductor lasers with inhomogeneous excitation". IEE PROC., vol.129, Pt.I, NO.4, AUGUST 1982

INFRARED NONLINEAR ABSORPTION AND PHOTOCONDUCTIVITY IN HgCdTe

Wang Weili Xing Qijiang Shi Shouxu
Physics Department, Peking University, China

There has been a resurgence of interest in the nonlinear optical properties of semiconductor materials for its potentical applications in optical bistability and the associated physical mechanism. Very large nonlinear refractive effects in narrow gap semiconductors have been investigated in the context of optical switching and phase conjugation[1]. This has led us to predict the existence of a series of intensity and frequency depended nonlinear absorption and photoconductivity observable at low incident power density in HgCdTe.

Using tunable laser photons with energy just slightly above the band gap of HgCdTe, the optical excited electrons vertically transit from the occupied states in the valence band to the empty states in the conduction band. For higher laser intensity optical transition into filled states is impossible and the corresponding absorption and photoconductivity appear saturation. That is dynamic Burstein-Moss effect. It is preferable to choose narrow gap semiconductors to observe these phenomena. Since small densities of states at bottom of conduction band are associated with small band gap. So, it is obvious that narrow gap semiconductor HgCdTe is a good candidate to realize these nonlinear effects in the infrared region.

The photon energy absorbed by semiconductors causes interband transition. It is obviously that the probability rate for transition and the corresponding absorption coefficient are related with the electron distribution function of initial states in valence band $f(I,E_v)$ and final states in conduction band $F(I,E_c)$. Including the laser intensity I and the angular frequency of laser line ω can be expressed by

$$\alpha(I,\omega)=\alpha_0(\omega)\left\{f(I,E_v)\left[1-f(I,E_c)\right]-f(I,E_c)\left[1-f(I,E_v)\right]\right\}$$
$$=\alpha_0(\omega)\left[f(I,E_v)-f(I,E_c)\right]$$

where α_0 is linear absorption coefficient at low incident power, electron distribution function is Fermi—Dirac distribution function. When photons excite additional electron-hole pairs and cause departure of the equalibrium condition, the unique Fermi level seperates into two quasi Fermi levels for electrons and holes respectively. Therefore with increasing laser intensity, electron and hole quasi Fermi levels approach the initial states E_v and final states E_c. We expect that that the occuped possibilities of E_c and E_v by electrons approach $\frac{1}{2}$, then electrons excited from valence band to conduction band are forbidden and semiconductors display the charactoristics of saturated optical absorption.

We have obtained the final expression which is similar to the relation between absorption coefficient and light intensity for two-level system with homogeneous broadening in the form[2]

$$\alpha(I,\omega)=\frac{\alpha_0(\omega)}{1+\frac{I(\omega)}{I_s(\omega)}}+\alpha_B$$

where α_B is a background linear absorption by excited electrons and holes, I_s is the saturation intensity parameter. Figure 1 gives the experimental results of intensity dependence of absorption coefficient at different laser lines. Also we can determine the linear absorption coefficients and saturation intensity parameter by our experiments as Tab. 1.

Tabulate 1

λ (μm)	λ^{-1} (cm^{-1})	α_0 (cm^{-1})	I_s (W/cm^2)
9.48	1055	230	19
9.68	1033	177	24
10.20	980	108	31
10.53	954	59	35
10.70	934	39	39

Semiconductor photoconductivity offers a reliable experimental means of extensively analizing semiconductor properties and basic physical mechanisms. Upon light illumination, nonequilibrium electron or hole concentration produced by intrinsic interband transition are determined by

$$\frac{dn(I,\omega)}{dt} = \frac{\alpha(I,\omega)I(\omega)}{\hbar\omega} - \frac{n(I,\omega)}{\tau_A}$$

Hence, laser excited electron or hole concentration is given as

$$n(I,\omega) = \alpha(I,\omega)I(\omega)\,\tau_A/\hbar\omega$$

For narrow gap semiconductors the dominant recombination is Auger process and strongly dependent on electron concentration. Then, the Auger recombination time is[3]

$$\tau_A = 2n_i^2\tau_{Ai}/n(I,\omega)^2$$

where n_i is the intrinsic carrier concentration and τ_{Ai} is the intrinsic Auger recombination time. The optical excited electron concentration can be expressed by

$$n(I,\omega)=\left[2n_i^2 \tau_{Ai}\alpha_o(\omega)/\hbar\omega\left(\frac{1}{I(\omega)}+\frac{1}{I_s(\omega)}\right)\right]^{1/3}$$

For weak optical excited condition $I \ll I_s$, the Auger recombination plays important role. The electron concentration appears $I^{1/3}$ dependence as

$$n(I,\omega)=\left[2n_i^2 \tau_{Ai}\alpha_0(\omega)I(\omega)/\hbar\omega\right]^{1/3}$$

For stronger optical excited condition $I \gg I_s$, we can obtain

$$n(I,\omega)=\left[2n_i^2 \tau_{Ai}\alpha_o(\omega)I_s(\omega)/\hbar\omega\right]^{1/3}$$

Hence, the electron concentration is saturated by both the Auger process and dynamic Burstein-Moss effect. We have used CW CO_2 laser lines as the light source to observe the saturable photoconductivity in $Hg_{0.8}Cd_{0.2}Te$ at 100K which is demonstrated in Figure 2. The intensity dependence of photoconductivity also appears nonlinear character.

REFERENCES

1. M.A.Khan et al., Opt. Lett., 6, 1560(1981).
2. Wang Wei-li, Chinese Physics, 7, 519(1987).
3. M.A.Kinch et al., J. Appl. Phys., 46, 1649(1973).

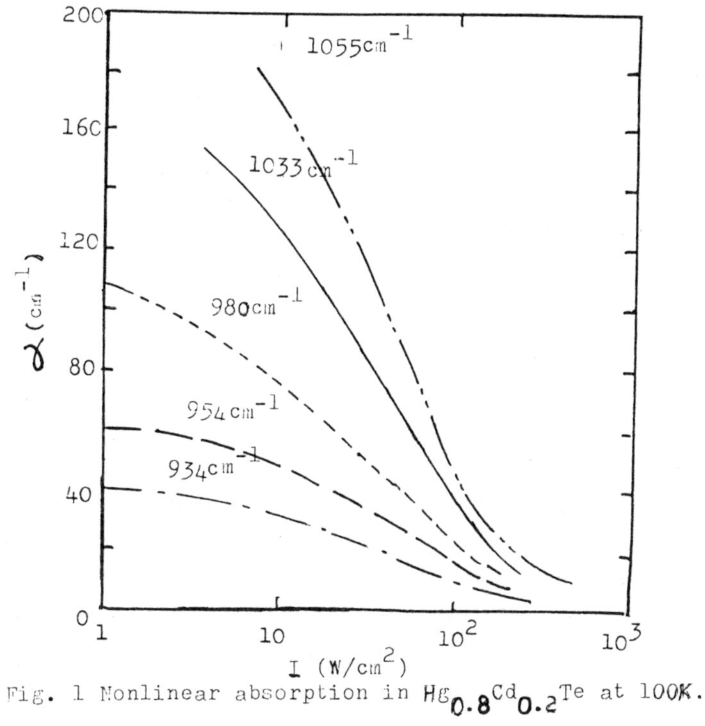

Fig. 1 Nonlinear absorption in $Hg_{0.8}Cd_{0.2}Te$ at 100K.

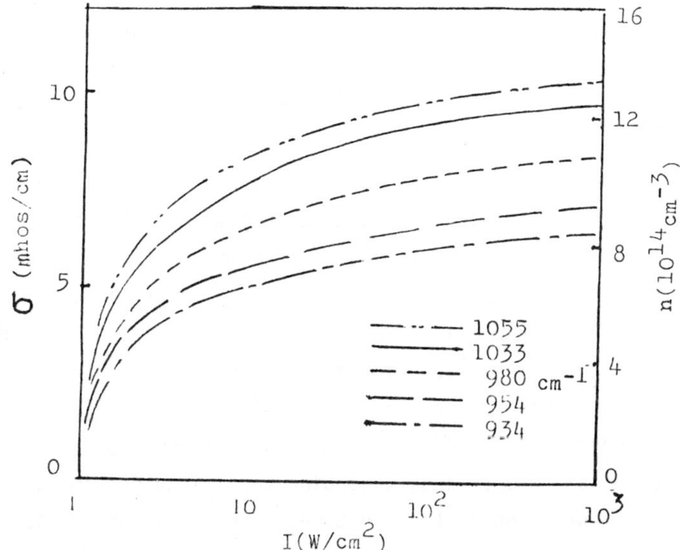

Fig. 2 Intensity dependences of nonlinear photoconductivity and excess electron concentration.

SWITCHING TIMES IN OPTICAL BISTABILITY: ANALYTICAL RESULTS.

Paul MANDEL

Université Libre de Bruxelles, Campus de la Plaine C.P. 231,
1050 Bruxelles, Belgium.

1. INTRODUCTION.

The main driving motivation behind current studies in optical bistability is the technological application to optical processing. This is clearly reflected in the gradual shift of the domain from fundamental to applied physics[1-5]. In assessing the potential applications of optical bistability, a recurring theme is the switching time. That is, the time it takes for a bistable system to switch from one stable state to another stable state. These two states are the physical supports of the logical levels "0" and "1". Clearly the switching time of a given bistable device is a crucial element in assessing the usefulness of optical processing, though taken alone it is of little significance. A more useful figure of merit is the switching time multiplied by the pixel density. The ultimate figure of merit will of course depend on the application considered.

The problem of switching times can be studied at two levels. In a microscopic theory, the physics of a particular device is modelled from first principles. The relevant mechanisms (electronic transitions, exciton or biexciton transitions, thermal gradients, surface properties,...) are described in terms of elementary interactions between matter and radiation. Macroscopic properties such as the transmission characteristics of the device result from an averaging procedure which relies on methods of statistical mechanics. Ideally, if such a program could be carried through, it would provide an expression of all properties, including the switching times, in terms of the fundamental constants of nature. Although progress has been reported in this direction (see chapter 2 of Ref.5) we are still quite far from the goal.

An alternative and rather complementary point of view is the macroscopic approach which will be followed in these notes. Since we only consider the class of bistable devices, let us take for granted that the device we study is characterized by a nonlinear characteristics of the type shown on Fig.1.

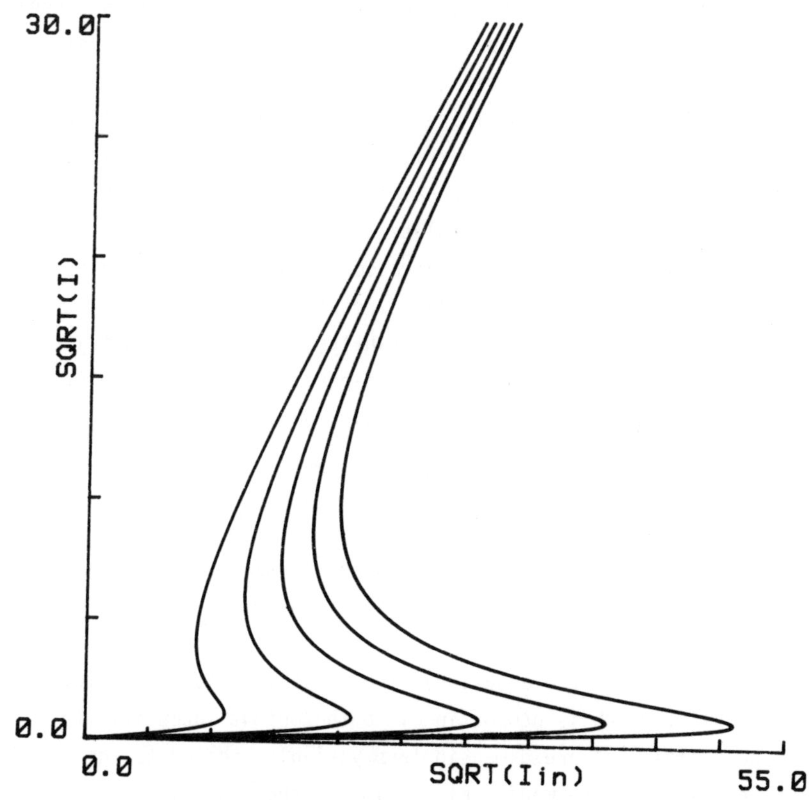

Figure 1. The one-parameter family of solutions of Eq.(1.1). From left to right the parameter C equals 10,20,30,40 and 50.

The implicit assumption of the macroscopic theory is that "universal" properties can be associated with the "universal" characteristics of Fig.1. This is justified, or rather motivated, by the fact that a large spectrum of very different physical mechanisms lead all to the same type of bistable response curves. And by extension one expects that some time-dependent properties will also be essentially model-independent.

Since we shall deal with a generic characteristics, it is essential that the mathematical model be expressed in terms of dimensionless variables: the physical time, the input intensity (or power) and the output intensity (or power) are scaled to characteristic microscopic time and intensities (or powers), respectively. As a result the decay rate of the field will be equal to one. Thus the model cannot predict the switching time in absolute units. Despite these restrictions on the applicability of the macroscopic approach, we shall see that it can lead to a number of usable results.

Our approach is based on a local analysis which we illustrate with an example. Figure 1 displays a one-parameter family of bistable curves corresponding to the characteristics equation

$$I_{in}=I\left(1+\frac{2C}{1+I}\right)^2 \qquad (1.1)$$

This equation derives from the analysis of the Maxwell-Bloch equations for a set of resonant two-level atoms in a ring cavity excited by an external input field of intensity I_{in} (see, e.g., Ref.6 for a derivation of this steady state equation). The variable I is the cavity field intensity which is proportional to the output intensity. In (1.1) the intensities are dimensionless variables scaled to the saturation intensity of the medium placed inside the ring cavity.

The relevant properties for our study are the coordinates of the limit points at which the upswitching or the downswitching process take place.

(i)In the limit $C \gg 1$ (fully developed hysteresis) we have for the up-switching point:

$I_{in,M}=C+O(1)$

$I_M=1+O(1/C)$

When $I_{in}=I_{in,M}$ the intensity I of the upper branch is given by $I=C+O(1)$.

(ii)Likewise in the same limit the downswitching point has coordinates

$I_{in,m}\cong\sqrt{8C}$

$I_m\cong\sqrt{2C}$

When $I_{in}=I_{in,m}$ the lower branch intensity is given by $I\cong(2/C)^{1/2}$.

To make a local analysis around the upswitching point, we introduce the scaling $I_{in}=C\mu$ and $I=x$. Then the coordinates of the upswitching point become $\mu=1$ and $x=1$. On the same scale, the point on the upper branch corresponding to $\mu=1$ has the coordinates $x=C>>1$ whereas the downswitching point is located at $\mu_m=\sqrt{8/C}<<1$ and $x_m=\sqrt{2C}$. Thus in first approximation we may replace the vicinity of the upswitching point by a parabola with a maximum at $(x,\mu)=(1,1)$ and consider that the upper branch is at an infinite distance (see figure 2).

Similarly to make a local analysis around the downswitching point we introduce the scaling $I_{in}=\mu\sqrt{8C}$ and $I=x\sqrt{2C}$. The downswitching point therefore has coordinates $(x,\mu)=(1,1)$. Furthermore when $\mu=1$ the corresponding point on the lower branch has the coordinate $x=(1/2C)^{1/2}<<1$ and the upswitching point is located at $\mu_M=(C/8)^{1/2}>>1$ and $x_M=(1/2C)^{1/2}<<1$. Thus in first approximation we may replace the vicinity of the downswitching point by a parabola with a maximum at $(x,\mu)=(1,1)$ and consider that the lower branch coincides with the $x=0$ axis (see figure 3).

It should be stressed that a similar analysis can be performed on generalizations of (1.1) which include detunings and transverse intensity profiles. Other mechanisms such as thermally-induced optical bistability[5,7] lead to curves similar to those of Fig.1 and there too a scaling can be established which leads to essentially similar conclusions.

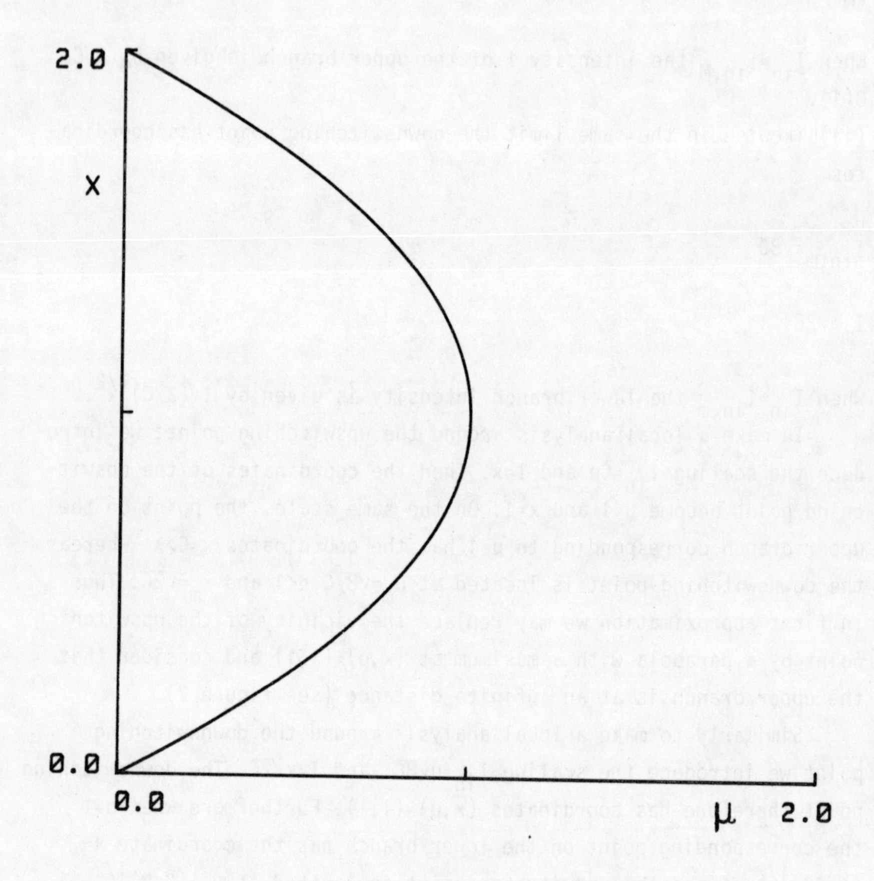

Figure 2. The vicinity of the upswitching point after rescaling.

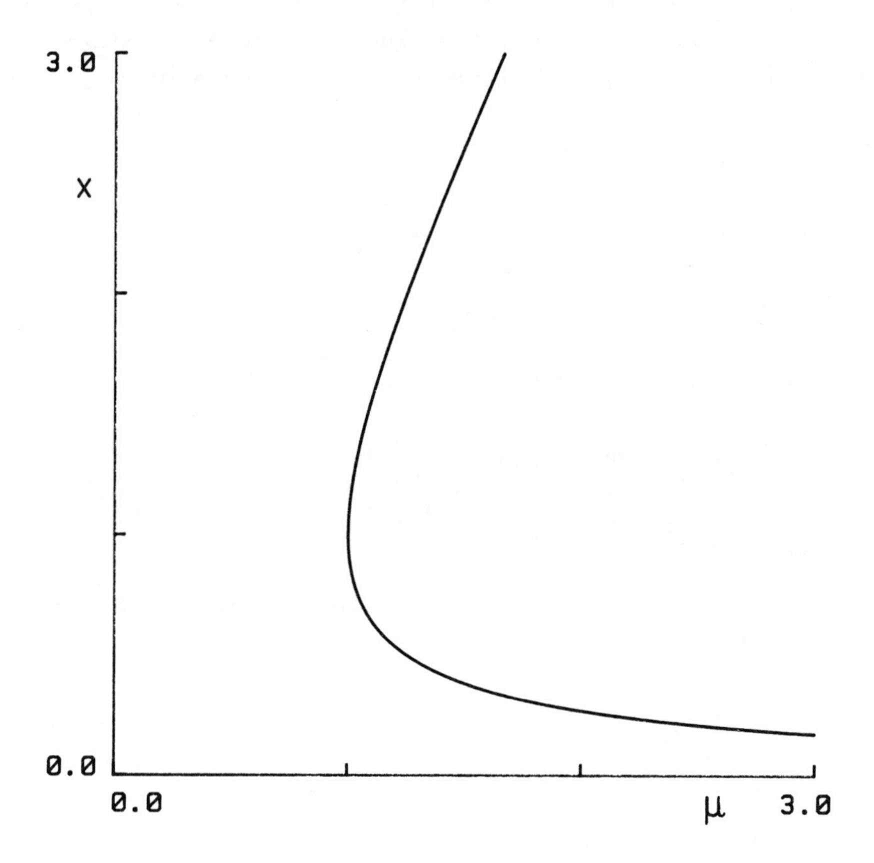

Figure 3. The vicinity of the downswitching point after rescaling.

2.UPSWITCHING DYNAMICS.

In the spirit of the scaling analysis outlined in the introduction, we shall model the dynamical behavior of an optically bistable device in the vicinity of the upswitching point by the equation[8]

$$dx/dt=x^2-2x+\mu \tag{2.1}$$

The steady solutions of (2.1) are (see figure 2):

$$x_{\pm}=1\pm\sqrt{1-\mu} \tag{2.2}$$

When $d\mu/dt=0$ it is easy to integrate (2.1) and to obtain

$$x(t)=\frac{x_-\{x(0)-x_+\}e^{\sqrt{1-\mu}t}-x_+\{x(0)-x_-\}e^{-\sqrt{1-\mu}t}}{\{x(0)-x_+\}e^{\sqrt{1-\mu}t}-\{x(0)-x_-\}e^{-\sqrt{1-\mu}t}} \tag{2.3}$$

Let us take $x(0)=x_++\alpha$ where α can be positive or negative. Then (2.3) can be written as

$$x(t)=x_- - \frac{\{x_+-x_-\}\{x_+-x_-+\alpha\}}{\alpha e^{2\sqrt{1-\mu}t}-\{x_+-x_-+\alpha\}} \tag{2.4}$$

This solution has the property

$$\lim_{t\to\infty} x(t)=\begin{cases} x_- & \text{if } \alpha<0 \\ x_+ & \text{if } \alpha=0 \\ \infty & \text{if } \alpha>0 \end{cases}$$

Thus for a given $\mu<1$ any initial condition below x_+ will eventually relax towards x_- whereas an initial condition above x_+ will diverge. This exact stability analysis indicates that x_- is a stable steady state whereas x_+ is unstable. Furthermore x_+ is a separatrix which limits two basins of attraction. We note that within the model description (2.1) a diverging solution corresponds to the initial stage of a transition to the upper branch which is located at infinity.

We now consider the response of the device to an input pulse which we take as rectangular. We shall analyze separately the response to long and to short pulses.

2A.Long input pulses.

Let us assume that the device is held initially in the stable state x_- corresponding to an input field $\mu_0<1$. At time t=0 the input field is suddenly increased to $\mu_1>1$. Thus we must solve (2.1) with the conditions:

$$t=0: \quad \mu=\mu_0<1, \quad x=x_-(\mu_0);$$
$$t>0: \quad \mu=\mu_1>1.$$

The solution of (2.1) in this case is

$$x(t)=\frac{(1-i\Omega)(\alpha+i\Omega)-(1+i\Omega)(\alpha-i\Omega)e^{-2i\Omega t}}{(\alpha+i\Omega)-(\alpha-i\Omega)e^{-2i\Omega t}} \tag{2.5}$$

with $\alpha=\sqrt{1-\mu_0}$ and $\Omega=\sqrt{\mu_1-1}$. This solution has zeros and poles. The first singularity is a pole at

$$e^{-2i\Omega t^*}=\frac{\alpha+i\Omega}{\alpha-i\Omega} \tag{2.6}$$

For t>t* the solution (2.5) still exists but is physically irrelevant. From (2.6) we have

$$\cos(2\Omega t^*)=\frac{\alpha^2-\Omega^2}{\alpha^2+\Omega^2}, \quad \sin(2\Omega t^*)=-\frac{2\alpha\Omega}{\alpha^2+\Omega^2} \tag{2.7}$$

In particular if $\mu_0=1$ (holding beam exactly at the limit point) the solution $t^*(\mu_0=1,\mu_1)=t^*(1,\mu_1)$ of (2.7) is

$$t^*(1,\mu_1)=\frac{\pi}{2\Omega} \tag{2.8}$$

Hence the jump transition to the upper branch takes place in a finite time which varies with the inverse square root of the pulse amplitude. Since $\mu_0=1$ the time $t^*(1,\mu_1)$ is the minimum jump duration under the long pulse conditions.

We now consider two extreme cases.

(i)Let the holding beam be very near (but below) the limit point and let the pulse amplitude be finite:

$$\mu_0=1-\varepsilon^2 \quad , \quad \mu_1-1=O(1) \tag{2.9}$$

In this limit the solution of (2.7) is

$$t^*(\mu_0,\mu_1)=t^*(1,\mu_1)+\frac{\varepsilon}{\Omega^2}+O(\varepsilon^2) \tag{2.10}$$

Since $t^*(1,\mu_1)$ is the jump duration, we can interpret $T=t^*(\mu_0,\mu_1)-t^*(1,\mu_1)$ as the destabilization time, i.e., the minimum pulse duration required to induce a transition between the lower and upper stable states. The total number of photons, N, delivered by the pulse to the device during the destabilization time T is

$$N=(\mu_1-\mu_0)T=\varepsilon\{1+\frac{2\varepsilon^2}{3\Omega^2}+\ldots\} \tag{2.11}$$

In first approximation $N=\varepsilon$ and therefore N is independent of the pulse amplitude: it only depends on the holding beam intensity. This means that in the operating conditions (2.9) all that matters is the <u>pulse area</u> rather than its amplitude or duration, separately.

(ii)Another extreme case is a holding beam far from the limit point and a pulse which brings the device barely above the limit point:

$$\mu_1=1+\varepsilon^2 \quad , \quad \mu_0-1=O(1) \tag{2.12}$$

Hence $\Omega=\varepsilon$ and the solution of (2.7) becomes

$$t^*(\mu_0,\mu_1)=\frac{\pi}{\varepsilon}-\frac{1}{\alpha}+O(\varepsilon) \tag{2.13}$$

In this unfavourable situation the switching time diverges. This is true for both the jump duration which equals $\pi/(2\varepsilon)$ and the destabilization time which equals $\pi/(2\varepsilon)$ $-1/\alpha$ $+O(\varepsilon)$. This divergence is a manifestation of critical slowing down. It expresses the fact that the dynamics of the device is ruled by a characteristic time proportional to the square root of the deviation of μ_1 from 1. This is clearly seen on (2.3) or (2.4). Since the limit point is defined by the coincidence of two solutions ($x_+=x_-$) it is a critical point. It can be proved[9] that critical points are characterized by an infinite relaxation time and therefore the vicinity of critical points has very large relaxation times.

2B.Short input pulses.

In this section we shall consider the effect of an input pulse of finite duration on the bistable device[10]. If the rectangular pulse duration is T and its amplitude is $\mu_1-\mu_0$, we have to solve (2.1) with the following conditions:

$t=0$: $\mu=\mu_0<1$, $x=x_-(\mu_0)$;

$0<t<T$: $\mu=\mu_1>1$;

$t>T$: $\mu=\mu_0<1$.

Under these constraints, (2.1) has the following solution:

$$x_1(t)=\frac{(1+i\Omega)(i\Omega-1+x_-)e^{-i\Omega t}+(1-i\Omega)(i\Omega+1-x_-)e^{i\Omega t}}{(i\Omega-1+x_-)e^{-i\Omega t}+(i\Omega+1-x_-)e^{i\Omega t}} \tag{2.15}$$

$$x_2(t)=\frac{x_+(X-x_-)e^{-x_+(t-T)}-x_-(X-x_+)e^{-x_-(t-T)}}{(X-x_-)e^{-x_+(t-T)}-(X-x_+)e^{-x_-(t-T)}} \tag{2.16}$$

where $x(t)=x_1(t)$ for $t<T$ and $x(t)=x_2(t)$ for $t>T$. In (2.15) Ω is defined as $\sqrt{\mu_1-1}$. In (2.16) $X\equiv x_1(T)$ whereas x_+ and x_- are the steady solutions (2.2) for $\mu=\mu_0<1$.

If the pulse is too short, the device will fall back on its initial state $x_-(\mu_0)$ for $t>T$ whereas for sufficiently long pulse duration, the jump transition to the upper branch (at infinity in this particular model) will be effected. Thus there must be a critical pulse duration, T^*, depending on both μ_0 and μ_1, which separates these two responses. It is easily seen from (2.16) that T^* is implicitly defined by the condition

$$X^*=x_1(T^*)=x_+ \tag{2.17}$$

from which it follows that $x_2(t)=x_+$ for all $t\geqq T^*$. Inserting (2.17) in (2.15) leads to

$$\cos(\Omega T^*)=\frac{\Omega^2-\alpha^2}{\Omega^2+\alpha^2}, \quad \sin(\Omega T^*)=\frac{2\alpha\Omega}{\Omega^2+\alpha^2} \tag{2.18}$$

where $\alpha=\sqrt{1-\mu_0}$. If the pulse duration T exceeds T^*, the solution (2.16) diverges in a finite time $t\uparrow$ given by

$$(X-x_-)e^{-x_+(t\uparrow-T)}=(X-x_+)e^{-x_-(t\uparrow-T)}$$

i.e.,

$$t\uparrow=T+\frac{1}{x_+-x_-}\ln\frac{X-x_-}{X-x_+} \tag{2.19}$$

In particular, when $T=T^*+\varepsilon$, $0<\varepsilon<<1$, we have $X=x_++\beta\varepsilon>x_+$ and the upswitching time diverges logarithmically

$$t\uparrow=\frac{1}{x_--x_+}\ln\varepsilon+O(1) \tag{2.20}$$

This logarithmic divergence of the switching time is much weaker than the algebraic divergence of the critical slowing down expressed by (2.13). Hence if μ is very near to unity, critical slowing down will always mask the logarithmic divergence.

We shall refer to the divergence (2.20) as slowing down since it does not arise from the vicinity of a critical point. Both critical and noncritical slowing down are universal effects associated with some kind of degeneracies. They are model-independent and can be expected whenever the control parameter brings the system in the vicinity of a critical point (critical slowing down) or in the vicinity of a separatrix (slowing down).

If the pulse duration T is less than the critical value T*, the solution (2.16) can be written as

$$x_2(t)=\frac{x_-+x_+e^{-F(t)}}{1+e^{-F(t)}} \tag{2.21}$$

$$F(t)=(x_+-x_-)(t-T)-\ln\frac{X-x_-}{x_+-X}$$

This solution describes the relaxation back to the initial state x_-. The relaxation time can be defined by the condition $F(t\downarrow)=1$, i.e.,

$$t\downarrow=T+\frac{1}{x_+-x_-}\{1+\ln\frac{X-x_-}{x_+-X}\} \tag{2.22}$$

If $T=T^*-\varepsilon$, $0<\varepsilon<<1$, we have $X=x_+-\beta\varepsilon<x_+$ and the relaxation time diverges logarithmically

$$t\downarrow=\frac{1}{x_--x_+}\ln\varepsilon\quad+0(1) \tag{2.23}$$

which again is a manifestation of noncritical slowing down.

Finally critical slowing down will occur if $\alpha=0(1)$ and $\Omega=\varepsilon$ in which case the solution of (2.18) is

$$T^* = \frac{\pi}{2\varepsilon} \tag{2.24}$$

In this occurence, the divergence of T^* in $t\uparrow$ and in $t\downarrow$ will completely mask the logarithmic divergence in Eqs.(2.19) and (2.23) since it is additive.

The results of this section are summarized on Figs. 4 to 6 which display the solution $x(t)$ of Eq.(2.1) versus time. On each curve the pulse duration, T, is indicated. The holding intensity, μ_0, the operating intensity, μ and the pulse amplitude $\mu-\mu_0$ are given by:

Figure 4:	$\mu_0=0.8$	$\mu=1.05$	$\mu-\mu_0=0.25$
Figure 5	$\mu_0=0.55$	$\mu=1.05$	$\mu-\mu_0=0.50$
Figure 6	$\mu_0=0.8$	$\mu=1.3$	$\mu-\mu_0=0.50$

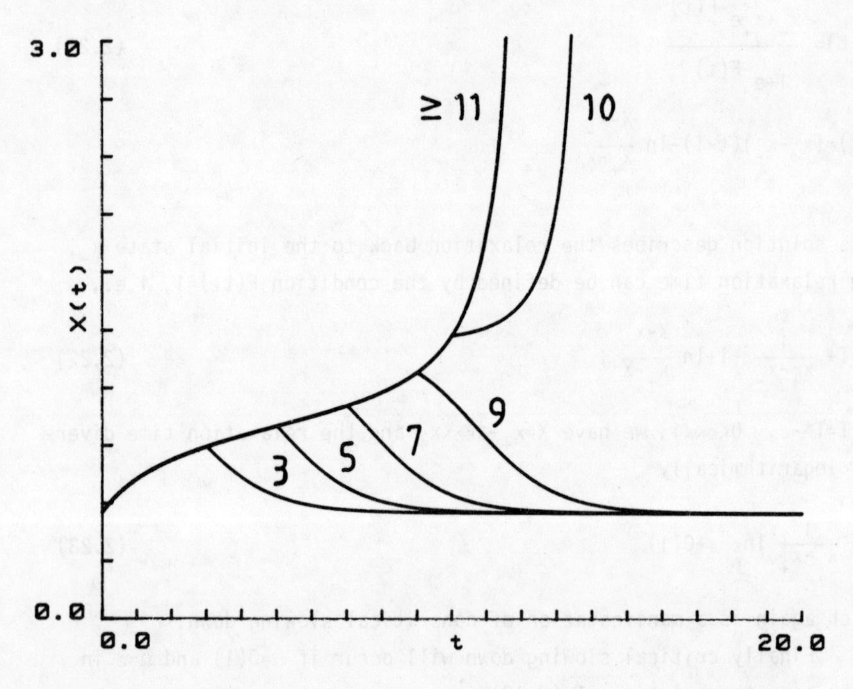

Figure 4. Solution $x(t)$ of Eq.(2.1) versus time for $\mu_0=0.8$ and $\mu=1.05$. The pulse duration is indicated on each curve.

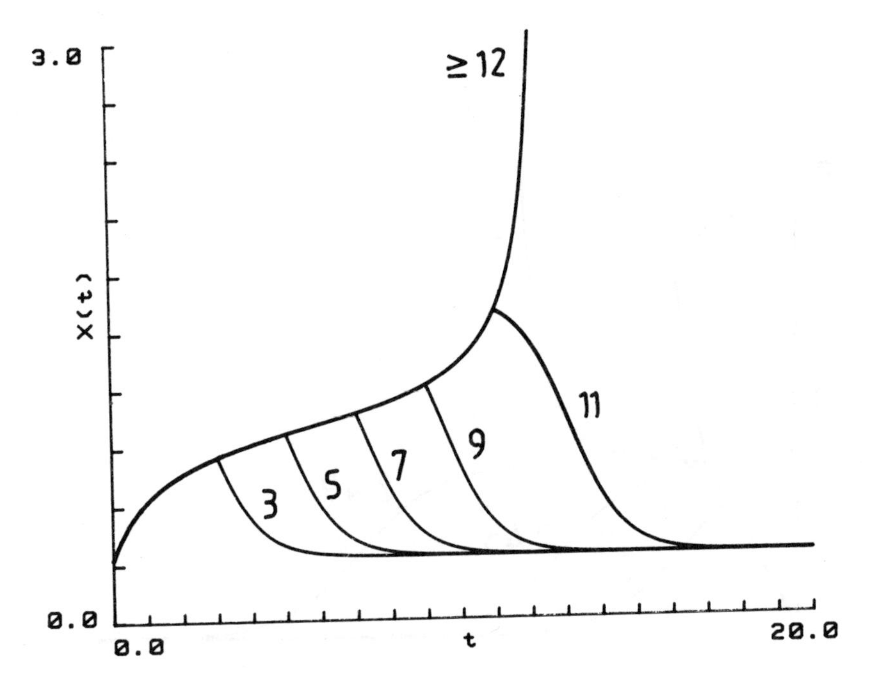

Figure 5. Solution x(t) of Eq.(2.1) versus time for μ_0=0.55 and μ=1.05. The pulse duration is indicated on each curve.

To make comparisons more easy, the same scale has been used for all three figures. Comparing Figs.4 and 5, we notice that when $\mu \cong 1$ the dynamics is hardly modified by a variation of the pulse amplitude. For long pulses (T>11) a variation of 100% in the pulse amplitude yields a variation of less than 10% in the switching time. If the operating intensity μ is removed from the vicinity of μ=1 as on Fig.6, then a much shorter switching time can be achieved. This is evident from comparing Figs.5 and 6 which correspond to identical pulse amplitudes. However too short pulses may be undesirable as well since they lead to (noncritical) slowing down as is manifest on Fig.6.

A similar analysis can be performed for the downswitching dynamics. A local analysis leads to the evolution equation

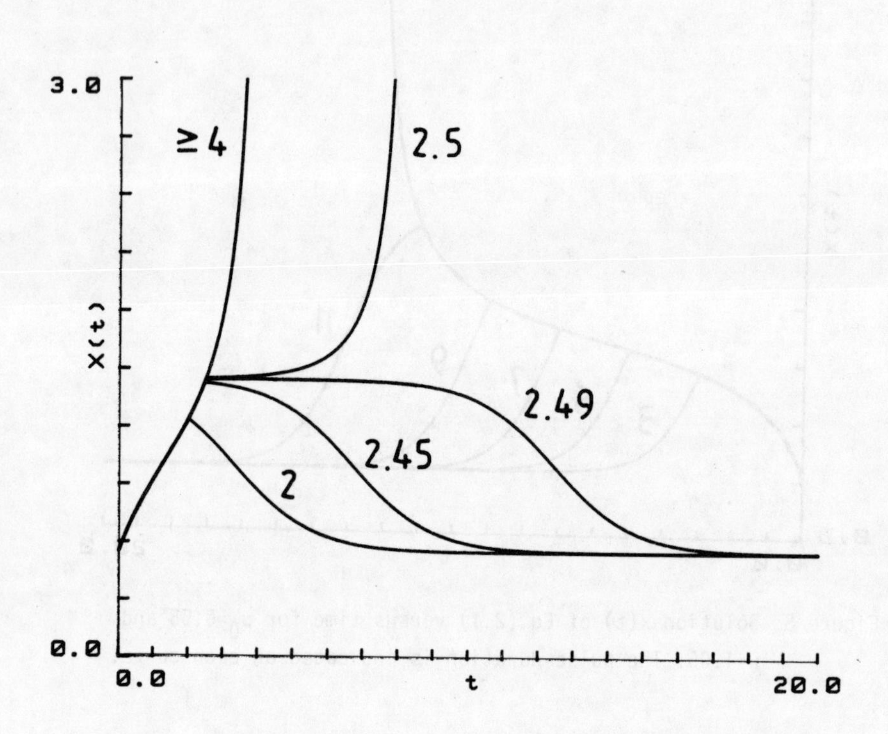

Figure 6. Solution x(t) of Eq.(2.1) versus time for μ_0=0.8 and μ=1.3. The pulse duration is indicated on each curve.

$$dx/dt=-x^2+2x\mu-1 \qquad (2.25)$$

whose steady solution is displayed on Fig.3. A study of (2.25) does not bring any new result and consequently will not be presented here.

3. COMPARISON WITH EXPERIMENTS.

Recently a number of experimental studies has been reported with the aim of testing the various results derived in section 2. The variety of bistable devices used in these reports also attests the generality of the theoretical analysis.

In Ref.10, a qualitative study of critical and noncritical slowing down has been reported on ZnSe nonlinear interference filters. These filters have a thickness of a few microns and can be addressed by Ar^+ lasers. Under favorable conditions[11] such filters can be switched in a few nanoseconds. On the contrary, if the operating conditions are such that slowing down or critical slowing down is present[10], the switching time can increase up to 50 milliseconds with a transient stabilization of the unstable middle branch of the hysteresis for as long as 15 milliseconds.

Quantitative measurements have been reported by the Lille (France) group under the direction of B. Macke. The experimental system is a 182 meter-long Fabry-Perot resonator filled with HCN at low pressure as a saturable absorber. This device is operated in the microwave domain ($\lambda \cong 3.5$mm). In a first paper[12] the pulse area scaling law (2.11) and the critical slowing down divergence of the switching time (2.13) have been confirmed. In a more recent paper[13] noncritical slowing down has also been observed and the $\ln\varepsilon$ divergence confirmed. As an example, Fig.7 shows a typical set of traces which record the output intensity versus time for different pulse durations. The operating conditions are such that critical slowing down is ruled out so that the lethargy time can be ascribed to noncritical slowing down. This is further confirmed by the transient stabilization of an intensity level corresponding to an unstable state and the fact that the lethargy time follows a logarithmic law.

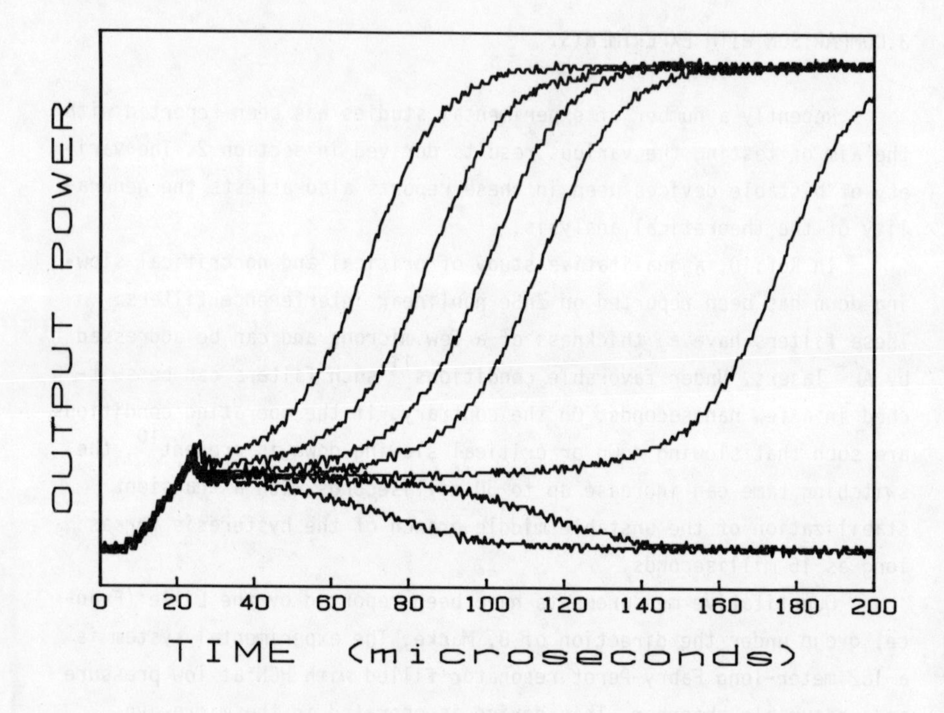

Figure 7. A typical set of output intensities versus time for the experimental set-up of the Lille group (see Ref.12). The upper and lower intensity levels of all pulses are constant. From left to right the pulse durations in μs are 18.939, 18.248, 17.921, 17.793, 17.730, 17.668 and 17.301.

ACKNOWLEDGMENTS

I wish to thank my friends and colleagues J.-Y.Bigot and A. Daunois at the Université de Strasbourg as well as B.Macke and B. Segard at the Université de Lille for their stimulating interest in this problem and for doing such nice experiments. Special thanks are due to the Lille group for providing the figure 7.
This research was supported by the Fonds National de la Recherche Scientifique (Belgium) and a grant from the European Commission.

REFERENCES.

1.Bowden,C.M., Ciftan,M. and Robl,H.R., eds. "Optical Bistabilty" (Plenum Press, New York) 1981.

2.Bowden,C.M., Gibbs,H.M. and McCall,S.L., eds. "Optical Bistabilty 2" (Plenum Press, New York) 1984.

3.Smith,S.D., Miller,A. and Wherrett,B.S., eds. "Optical Bistability, Dynamical Nonlinearity and Photonic Logic" (The Royal Society, London) 1985.

4.Gibbs,H.M., Mandel,P., Peyghambarian,N. and Smith,S.D., eds. "Optical Bistability 3" (Springer Proceedings in Physics, vol.8, Heidelberg) 1986.

5.Mandel,P., Smith,S.D. and Wherrett,B.S., eds. "From Optical Bistability Towards Optical Computing: The EJOB Project" (North-Holland, Amsterdam) 1987.

6.Lugiato,L.A. "Theory of Optical Bistability" in Progress in Optics, vol. XXI, pp71-216 (North-Holland, Amsterdam)1984.

7.Gibbs,H.M., "Optical Bistability: Controlling Light With Light" (Academic Press, New York) 1985.

8.Mandel,P., Optics Commun. 55 (1985)293.

9.Mandel,P., "Dynamic versus static stability" in Frontiers in Quantum Optics, E.R.Pike and S.Sarkar eds. (Adam Hilger, Bristol) pp430-452, 1986.

10.Bigot,J.-Y., Daunois,A. and Mandel,P., Phys.Lett. (in press).

11.Bigot,J.-Y., Daunois,A., Leonelli,R., Sence,M., Mathew,J.G.H., Smith,S.D. and Walker,A.C., Appl.Phys.Lett. 49 (1986) 844.

12.Segard,B., Zemmouri,J. and Macke,B., Optics Commun. 60 (1986)323.

12.Segard,B., Zemmouri,J. and Macke,B., Optics Commun. (in press).

EIGENVALUES OF THE QUANTUM-FOKKER-PLANCK EQUATION FOR DISPERSIVE OPTICAL BISTATBILITY

H. Risken, K. Vogel
Abteilung für Theoretische Physik, Universität Ulm
D-7900 Ulm, Federal Republic of Germany

1 INTRODUCTION

Optical bistability has become an important field in quantum optics, see for instance [1-4] for reviews. Such systems may exist in either of two macroscopic states. Quantum tunneling leads to a transition between these two states and thus, in the absence of any other fluctuations, determines the ultimate stability of these states. For absorptive optical bistability it was shown recently that the tunneling rate can be reduced by coupling to a squeezed vacuum [5]. A fully quantum mechanical treatment of optical bistability requires the solution of the master equation, i. e. the equation of motion for the density operator. For the model of Drummond and Walls (DW) [6] describing dispersive optical bistability we solve this master equation in the stationary state and we calculate some of the lowest nonzero eigenvalues. These eigenvalues then determine the tunneling rates. The DW model has the advantage that only the operators of the cavity light mode enter in the equation of motion of the density operator. Besides this simplicity it is a nonlinear and nontrivial model.

In order to solve the master equation for the density operator we first transform the master equation for the density operator into a Fokker-Planck equation for the Q-function [7], i.e for the expectation value of the density operator with respect to the coherent state $|\alpha\rangle$. Because the diffusion matrix is not positive definite or semidefinite it is not an ordinary FPE which can be interpreted as describing the Brownian motion of a particle in a suitable potential. We have termed such a FPE a quantum-Fokker-Planck equation (QFPE). For a non positive definite diffusion matrix we may still have a stable stationary solution. An illustrative example is the equation

$$\frac{\partial W}{\partial t} = \left[\frac{\partial}{\partial x_1}(x_1-\alpha x_2) + \frac{\partial}{\partial x_2}(x_2+\alpha x_1) + \frac{\partial^2}{\partial x_1^2} - q\frac{\partial^2}{\partial x_2^2} \right] W \;, \tag{1.1}$$

which has a stable stationary solution if the conditions

$$q < 1 \;, \quad (1+q)^2/(1-q)^2 < 1+\alpha^2 \tag{1.2}$$

are fulfilled. A stationary solution of the master equation for the density operator for $n_{th} = 0$, i.e. without any thermal fluctuations was already obtained by DW using a complex P-representation of the density operator. The complex P-function as well as the positive P-function have been introduced and further investigated by Gardiner [8]. As was shown by DW expectation values of the light field operators can be expressed in terms of generalized Gauss hypergeometric series. By applying the matrix continued fraction (MCF) method for solving two variable FPEs [9] we obtain the stationary solutions as well as the lowest nonzero real eigenvalues and some other low real eigenvalues of the QFPE for the Q-function [7,10]. The lowest nonzero eigenvalue is connected with the tunneling rate [11], which, as already mentioned, determines the ultimate stability of the system. The MCF method is not only applicable for pure quantum fluctuations where the number of thermal quanta n_{th} is zero. It is also applicable to the case $n_{th} > 0$ where the detailed balance condition for the complex P-function is no longer valid and therefore a stationary solution is hard to obtain.

The present paper is organized as follows. In Chap. 2 we present the DW model as well as the classical equation of motion without fluctuations. In Chap. 3 we derive the QFPE for the P- and Q function and outline in Chap. 4 the MCF method for solving it. Finanlly in Chap. 5 we calculate some lowest real eigenvalues. In particular results are shown for small cavity damping constants.

2 MODEL AND BASIC EQUATIONS

By expanding the polarization up to third order, by including a coherent classical driving field, by adding losses due to cavity damping and by making the rotating-wave approximation DW obtained a master equation for the density operator of the light field inside the cavity. In a slightly different notation this master equation takes the form

$$\dot{\rho} = -i[H,\rho] + \kappa L_{ir}[\rho] \ , \tag{2.1}$$

where H and L_{ir} are given by

$$H = -\Omega a^{\dagger}a + \chi a^{\dagger 2}a^{2} - F(a+a^{\dagger}) \tag{2.2}$$

$$L_{ir}[\rho] = 2a\rho a^{\dagger} - \rho a^{\dagger}a - a^{\dagger}a\rho + 2n_{th}[[a,\rho],a^{\dagger}] \ . \tag{2.3}$$

Here $\Omega = \omega_l - \omega_c$ is the difference between the frequency of the classical driving field F and the cavity frequency, χ is the imaginary part of the third order susceptibility, κ is the cavity damping constant, n_{th} is the number of thermal quanta and $a^{\dagger}$ and a are the creation and annihilation operators for the light field inside the cavity.

From (2.1) we obtain for the complex amplitude

$$\alpha = \mathrm{Tr}(a\rho) \ , \qquad \alpha^{*} = \mathrm{Tr}(\alpha^{\dagger}\rho) \tag{2.4}$$

$$\dot{\alpha} = i\Omega\alpha - \kappa\alpha - 2i\chi\,\mathrm{Tr}(a^{\dagger}a^{2}\rho) + iF \ . \tag{2.5}$$

By replacing $\mathrm{Tr}(a^{\dagger}a^{2}\rho)$ in terms of the expectation value (2.4), i. e. by $\alpha^{*}\alpha^{2}$ we arrive at the classical equation whithout fluctuations

$$\dot{\alpha} = [i\Omega - \kappa - 2i\chi\alpha^{*}\alpha]\alpha + iF \ . \tag{2.6}$$

By using the normalized time $\tilde{t}$, amplitude $\tilde{\alpha}$, intensity $\tilde{I} = \tilde{\alpha}^{*}\tilde{\alpha}$, damping constant $\tilde{\kappa}$ and driving field $\tilde{F}$ defined by $(\Omega > 0)$

$$\tilde{t} = \Omega t, \quad \tilde{\alpha} = \sqrt{\chi/\Omega}\,\alpha, \quad \tilde{I} = (\chi/\Omega)\,I, \quad \tilde{\kappa} = \kappa/\Omega, \quad \tilde{F} = \frac{1}{\Omega}\sqrt{\chi/\Omega}\,F \tag{2.7}$$

(2.6) is transformed to the normalized form

$$d\tilde{\alpha}/d\tilde{t} = [i(1-2\tilde{\alpha}^*\tilde{\alpha}) - \tilde{\kappa}]\tilde{\alpha} + i\tilde{F}. \quad (2.8)$$

It follows from (2.8) that the following connection between $|\tilde{F}|$ and $\tilde{I}$ is valid for the stationary state ($d\tilde{\alpha}/d\tilde{t} = 0$, see also [6])

$$|\tilde{F}| = \sqrt{\tilde{I}[\tilde{\kappa}^2 + (1-2\tilde{I})^2]} \quad . \quad (2.9)$$

According to DW the stationary solution is unstable between the turning points, see [10,11]. Bistabilty occurs if we have turnig points, i. e. for

$$\tilde{\kappa} < 1/\sqrt{3}, \quad |\tilde{F}| < \sqrt{1 + 9\tilde{\kappa}^2 + (1-3\tilde{\kappa}^2)^{3/2}}\Big/(3\sqrt{3}) \ . \quad (2.10)$$

3 QUANTUM-FOKKER-PLANCK EQUATION

Any normally or antinormally expectation value of the light field operators a and $a^\dagger$ may be obtained from the characteristic functions

$$\tilde{P}(\beta) = \tilde{P}^+(\beta) = \mathrm{Tr}\{e^{i\beta^* a^\dagger} e^{i\beta a} \rho\} ,$$

$$\tilde{Q}(\beta) = \tilde{P}^-(\beta) = \mathrm{Tr}\{e^{i\beta a} e^{i\beta^* a^\dagger} \rho\} \quad (3.1)$$

by appropriate differentiation with respect to β and β^*. The Fourier-transforms of these characteristic functions

$$P^\pm(\alpha) = \pi^{-2} \int e^{-i\alpha\beta - i\alpha^*\beta^*} \tilde{P}^\pm(\beta)\, d^2\beta \quad (3.2)$$

are the Glauber-Sudarshan P-function and the Q-function, see for instance [12]. Because the equations of motion for the P- and the Q-function derived later on differ only by ± signs, we have used in (3.1,3.2) and will use later on the notation

$$P(\alpha) = P^+(\alpha) \ ; \quad Q(\alpha) = P^-(\alpha) \ . \quad (3.3)$$

The Q-function can be expressed by the density operator according to

$$Q(\alpha) = P^{-}(\alpha) = \langle\alpha|\rho|\alpha\rangle/\pi\,, \qquad |\alpha\rangle = \text{coherent state} \tag{3.4}$$

whereas the density operator ρ itself can be expressed by the Glauber-Sudarshan P-function [13,14] by the relation

$$\rho = \int |\alpha\rangle\langle\alpha|\, P(\alpha)\, d^2\alpha\,. \tag{3.5}$$

Normally and antinormally ordered expectation values are obtained from the P- and Q- function by integration

$$\langle a^{\dagger n} a^m\rangle = \int \alpha^{*n}\alpha^m P(\alpha)\, d^2\alpha\,; \qquad \langle a^m a^{\dagger n}\rangle = \int \alpha^{*n}\alpha^m Q(\alpha)\, d^2\alpha\,. \tag{3.6}$$

Because of squeezing [15] the P-function does not exist in general. As will be seen later on the expansion coefficients of the P-function into a complete set, however, do exist. (Also its Fourier transform $\tilde{P}(\beta)$ does exist.) In order to derive an equation for these expansion coefficients we may nevertheless use the equation of motion for the P-function.

From the master equation (2.1) the following equation for the $P=P^{+}$- and $Q=P^{-}$-function was derived [6]

$$\begin{aligned}\frac{\partial P^{\pm}}{\partial t} = &- \frac{\partial}{\partial\alpha}\left(-\kappa\alpha + i\Omega\alpha + 2i(1\mp 1)\chi\alpha - 2i\chi\alpha^2\alpha^* + iF\right)P^{\pm} \\ &- \frac{\partial}{\partial\alpha^*}\left(-\kappa\alpha^* - i\Omega\alpha^* - 2i(1\mp 1)\chi\alpha^* + 2i\chi\alpha^{*2}\alpha - iF\right)P^{\pm} \\ &\mp i\chi\frac{\partial^2}{\partial\alpha^2}\alpha^2 P^{\pm} + 2\kappa\left(n_{th} + \frac{1}{2} \mp \frac{1}{2}\right)\frac{\partial^2 P^{\pm}}{\partial\alpha^*\partial\alpha} \pm i\chi\frac{\partial^2}{\partial\alpha^{*2}}\alpha^{*2}P^{\pm}\,.\end{aligned} \tag{3.7}$$

The upper signs are valid for the P-function, the lower ones for the Q-function. If we use the intensity I and the phase ϕ defined by

$$I = \alpha^*\alpha, \qquad \alpha = \sqrt{I}\, e^{i\phi} \tag{3.8}$$

(3.7) is transformed to

$$\frac{\partial P^{\pm}}{\partial t} = \Big\{ - \frac{\partial}{\partial I}\Big(-2\kappa I + \kappa(2n_{th} + 1 \mp 1) + 2F\sqrt{I}\sin\phi\Big)$$

$$- \frac{\partial}{\partial \phi}\Big(\Omega - 2\chi(I-1) \mp \chi + \frac{F}{\sqrt{I}}\cos\phi\Big)$$

$$+ \kappa(2n_{th} + 1 \mp 1)\frac{\partial^2}{\partial I^2}I \mp 2\chi\frac{\partial^2}{\partial\phi\partial I}I + \frac{\kappa}{2I}(n_{th} + \frac{1}{2} \mp \frac{1}{2})\frac{\partial^2}{\partial\phi^2}\Big\} P^{\pm} . \tag{3.9}$$

It is easily derived from (3.9) that the diffusion matrix is not positive definite if the intensity is large enough, i. e. for

$$I = \alpha^*\alpha > \frac{\kappa}{\chi}(n_{th} + \frac{1}{2} \mp \frac{1}{2}) . \tag{3.10}$$

Because the diffusion matrix is not positive definte or positive semidefinite everywhere, (3.7,9) cannot be interpreted as describing the Brownian motion of a particle in a suitable potential and therefore no simple simulation of (3.7,9) is possible. For this reason (3.7,9) was termed quantum-Fokker-Planck equation (QFPE). By doubling the phase space [8] it may be possible to derive a FPE with a positive definite diffusion matrix and a simulation is then possible [16]. (By addding Langevin noise forces to (2.6) Graham and Schenzle [17] and Haug et al. [18] obtained a FPE for dispersive optical bistability with a positive definite diffusion matrix. It is similar to (3.7) for $n_{th} \gg 1$.)

4 SOLUTION IN TERMS OF MATRIX CONTINUED FRACTIONS

For the $P=P^+$- and $Q=P^-$-function we use the real expansion

$$P^{\pm}(I,\phi) = \sum_{m=0}^{\infty} \Big\{ a_m^0 \exp(-I/I_s)\, L_m^0(I/I_s)$$

$$+ \sum_{n=1}^{\infty} \frac{2}{\sqrt{n!}}\big(a_m^n \cos n\phi - b_m^n \sin n\phi\big) \exp(-I/I_s)\,(I/I_s)^{n/2} L_m^n(I/I_s)\Big\} , \tag{4.1}$$

where $L_m^n(I/I_s)$ are the generalized Laguerre polynomials. (An expansion of this type was already used in [18,19].) The scaling intensity I_s is arbitrary but will be chosen such to achieve good numerical convergence of the expansion (4.1). The factor $1/\sqrt{n!}$ was added to reduce the numerical errors for the inversion of the matrices. In the next step we insert (4.1) into (3.9) and obtain the following recurrence relations for the expansion coefficients

$$\dot{a}_m^0 = \left(2F/\sqrt{I_s}\right) m\, b_{m-1}^1 - \frac{2\kappa}{I_s}\left(n_{th} + \frac{1}{2} \mp \frac{1}{2} - I_s\right) m\, a_{m-1}^0 - 2\kappa m\, a_m^0$$

$$\dot{a}_m^n = \left(F/\sqrt{I_s(n+1)}\right) m\, b_{m-1}^{n+1} + F\sqrt{n/I_s}\, b_m^{n-1} - \frac{2\kappa}{I_s}\left(n_{th} + \frac{1}{2} \mp \frac{1}{2} - I_s\right) a_{m-1}^n$$

$$- 2\kappa(m+n/2) a_m^n - n\left\{-\Omega + \chi(2I_s \pm 1)(2m+n) + \chi(2I_s - 2 \pm 1)\right\} b_m^n$$

$$+ 2nm\chi(I_s \pm 1)\, b_{m-1}^n + \frac{2\chi}{I_s} n(m+n+1)\, b_{m+1}^n$$

$$\dot{b}_m^n = -\left(F/\sqrt{I_s(n+1)}\right) m\, a_{m-1}^{n+1} - F\sqrt{n/I_s}\, a_m^{n-1} - \frac{2\kappa}{I_s}\left(n_{th} + \frac{1}{2} \mp \frac{1}{2} - I_s\right) m\, b_{m-1}^n$$

$$- 2\kappa(m+n/2) b_m^n + n\left\{-\Omega + \chi(2I_s \pm 1)(2m+n) + \chi(2I_s - 2 \pm 1)\right\} a_m^n$$

$$- 2nm\chi(I_s \pm 1)\, a_{m-1}^n - \frac{2\chi}{I_s} n(m+n+1)\, a_{m+1}^n \,. \qquad (4.2)$$

It should be noted that the normalization

$$\int P(\alpha) d^2\alpha = \int Q(\alpha) d^2\alpha = \int P^{\pm}(\alpha) d^2\alpha = 1 \qquad (4.3)$$

requries $a_0^0 = 1/(I_s\pi)$ in both cases. If squeezing occurs the expansion (4.1) for the P-function does not exist, though the expansion coefficients of the P-function do exist. The equation of motion for the expansion coefficients of the P-function (4.2) determine these coefficients. In principle, the P-function and the corresponding QFPE can be avoided by only using the the master equation (2.1) and the connection between moments and expansion coefficients for the derivation of (4.2).

In (4.2) only coefficients with adjacent indices are coupled. By introducing column vectors of the form

$$\mathbf{c}_m = (a_m^0,\ a_m^1,\ b_m^1,\ a_m^2,\ b_m^2,\ldots) \tag{4.4}$$

we can cast the recurrence relations (4.2) into the tridiagonal vector recurrence relation

$$\dot{\mathbf{c}}_m = \mathbf{Q}_m^- \mathbf{c}_{m-1} + \mathbf{Q}_m \mathbf{c}_m + \mathbf{Q}_m^+ \mathbf{c}_{m+1}\ , \tag{4.5}$$

where $\mathbf{Q}_m^\pm$ and $\mathbf{Q}_m$ are matrices following from (4.2). By investigating the Brownian motion problem in tilted periodic potentials Vollmer and one of us had derived tridiagonal vector recurrence relations of the form (4.5) [20]. In these references the stationary solution, eigenvalues, eigenfunctions as well as some other instationary solutions of an equation of the form (4.5) have been obtained by calculating appropriate matrix continued fractions (MCF), see also [9] Chaps. 9 and 11 for a review. The same MCF method can be applied to (4.5). In order to calculate the matrix continued fractions, the expansion (4.1) has to be truncated at a large but finite index $n = L$, so that the matrices $\mathbf{Q}$, $\mathbf{Q}^\pm$ in (4.5) have the dimension $(2L+1)*(2L+1)$. Furthermore, the infinite continued fractions have to be replaced by their Mth approximants. This means that the expansion (4.1) is also truncated at the index $m = M$. The truncation indices L and M have to be chosen such that the final results do not change within a given accuracy if L and M is increased.

5 RESULTS

Here we confine ourselves to the investigation of the eigenvalues of the QFPEs (3.7,9) which determine the tunneling rates and therefore the stability of the system. Some stationary results can be found in [6,7,10,11]. In Fig. 1 we have plotted some of the lowest nonzero eigenvalues calculated from (4.2) with the MCF method. As it is clearly seen the lowest nonzero eigenvalue becomes very small in the middle of the bistable region. As explained in [11] this lowest

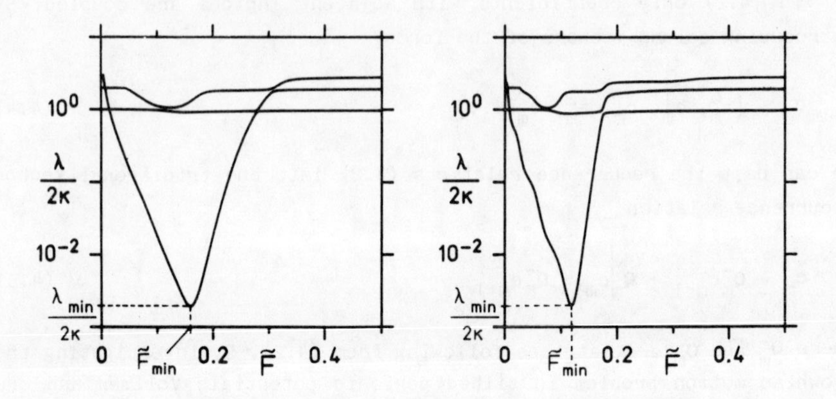

Fig. 1: Plot of the lowest real nonzero eigenvalues divided by 2κ in a logarithmic scale as a function of $\tilde{F}$ for $\tilde{\kappa} = 0.001$, $n_{th} = 0$ and $\Omega/\chi = 9.5$ (a), 10.0 (b)

eigenvalue determines the tunneling rate. Comparing Fig. 1a to Fig 1b one realizes that the curves for the first nonvanishing eigenvalue changes appreciably if the parameter Ω/χ is slightly increased from 9.5 to 10.0. In order to investigate this behavior we calculated the field $\tilde{F}_{min}$ where the first nonvanishing eigenvalue has a minimum and the corresponding eigenvalue λ_{min} as a function of the parameter Ω/χ. The results are shown in Fig. 2. As one can clearly see the field $\tilde{F}_{min}$ reaches very low values at integer values of Ω/χ and increases very sharply by moving away from integer values. The corresponding minimum eigenvalue decreases roughly exponentially for increasing values of Ω/χ. At integer values of Ω/χ, however, the eigenvalue has a peak leading to somewhat larger values at and near integer values for Ω/χ. This oscillating behavior disappears for larger damping constants.

It seems that this unexpected behavior of the minimum eigenvalue is a typical quantum effect stemming from the discrete energy levels of the Hamilton operator (2.2). (For a discussion of the energy levels of this operator see [11].) By increasing Ω/χ a new eigenstate between the state which evolves from the vacuum state at F=0 and the minimum energy state occurs at integer values of Ω/χ. In [11] eigenvalues have been obtained by using an appropriate Pauli master equation. For noninteger

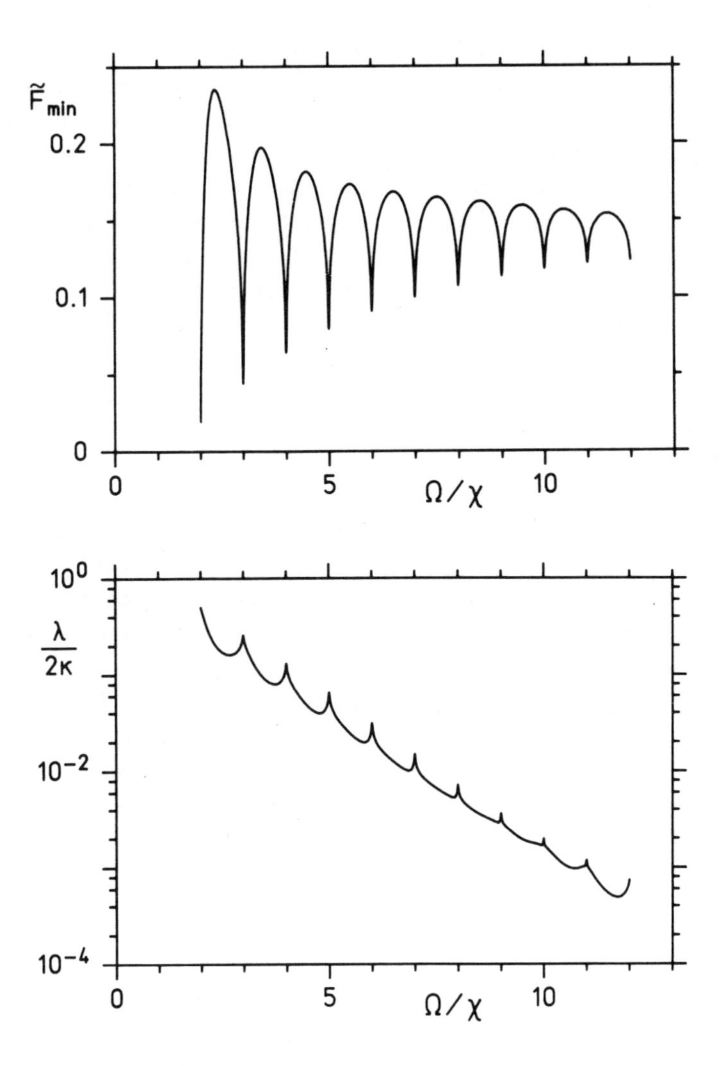

Fig. 2: The field $\tilde{F}_{min}$ where the first nonvanishing eigenvalue has a minimum (a) and the corresponding minimum eigenvalue divided by 2κ (b) as a function of Ω/χ for $\tilde{\kappa} = 0.001$, $n_{th} = 0$

Ω/χ these eigenvalues agree with the ones obtained by the MCF method for small κ. (For integer or near integer values of Ω/χ the Pauli master equation used in [11] does not work. Here also nondiagonal elements of ρ have to be taken into account.) It should be noted that we get the same eigenvalues for both P-and Q-function. The lowest nonzero eigenvalue is connected with the escape rate [11]. Thus also the decay rate agrees for both functions. This is of course an essential feature of a fully quantum mechanical theory. As remarked by Drummond [21] in previous calculations on absorptive optical bistability approximations were made which are not consistent with fully quantum mechanical calculations. These calculations lead to tunneling rates which - depending on the representation used - differ by orders of magnitude.

7 CONCLUSION

In conclusion we have solved the QFPE for the Q- as well as for the P-function with a non positive definite diffusion matrix describing optical bistability with the matrix continued fraction method. In particular we investigated the tunneling times by calculating some lowest nonzero eigenvalues. The eigenvalues of the equation for the P- and the Q-function agree very accurately. The lowest nonzero real eigenvalue has a minimum approximately in the middle of the bistable region. An appreciable oscillating variation of the location of this minimum and the eigenvalue itself with the ratio of the detuning to the parameter of the nonlinear susceptibility was found for small cavity damping.

REFERENCES

1 C. M. Bowden, M. Cliftan, H. R. Robl (Eds.): Optical bistability, Plenum Press, New York (1981)

2 R. Bonifacio (Ed.): Dissipative Systems in Quantum Optics, Topics in Current Physics, Vol. 27, Springer, Berlin (1982)

3 L. A. Lugiato, In: Progress in Optics XXI, Ed. E. Wolf, page 69, North-Holland, Amsterdam (1984)

4 J. C. Englund, R. R. Snapp, W. C. Schieve, In: Progress in Optics XXI, Ed. E. Wolf, page 355, North-Holland, Amsterdam (1984)

5 C. M. Savage, D. F. Walls, Phys. Rev. Lett. **57**, 2164 (1986)

6 P. D. Drummond, D. F. Walls, J. Phys. **A13**, 725 (1980)

7 K. Vogel, H. Risken, Optics Comm. **62**, 45 (1987)

8 C. W. Gardiner, Handbook of Stochastic Methods, Springer Series in Synergetics, Vol. 13, 2nd ed., Springer, Berlin (1985)

9 H. Risken, The Fokker-Planck Equation, Springer Series in Synergetics, Vol. 18, Springer, Berlin (1984)

10 H. Risken, K. Vogel, In: Fundamentals of Quantum Optics II, Ed. F. Ehlotzky, page 225, Springer, Berlin 1987

11 H.Risken, C. Savage, F. Haake, D. F. Walls, Phys. Rev. **A35** 1729 (1987)

12 H. Haken, Laser Theory in Encyclopedia of Physics, Vol. XXV/2c, Ed. S. Flügge, Springer, Berlin (1970)

13 R. J. Glauber, Phys. Rev. **131**, 2766 (1963)

14 E. C. G. Sudarshan, Phys. Rev. Lett. **10**, 277 (1963)

15 D. F. Walls, Nature **306**, 141 (1983)

16 H. Dörfle, A. Schenzle, Z. Phys. **B65**, 113 (1986)

17 R. Graham, A. Schenzle, Phys. Rev. **A23**, 1302 (1981)

18 H. Haug, S. W. Koch, R. Neumann, H. E. Schmidt, Z. Phys. **B49**, 79 (1982)

19 H. Risken, H. D. Vollmer, Z. Phys. **B39**, 89 and 339 (1980)

20 H. Risken, H. D. Vollmer, Z. Phys. **B33**, 297 (1979)
H. D. Vollmer, H. Risken, Z. Phys. **B34**, 313 (1979)
H. D. Vollmer, H. Risken, Physica **110A**, 106 (1982)
H. Risken, H. D. Vollmer, Mol. Phys **46**, 555 (1982)

21 P. D. Drummond, Phys. Rev. **A33**, 4462 (1986)

Bistable Three-Photon Resonance in Semiconductors

A. E. Kaplan and Y. J. Ding

Department of Electrical and Computer Engineering
The Johns Hopkins University
Baltimore, MD 21218, USA

Abstract

The feasibility of a bistable pseudorelativistic three-photon cyclotron resonance in narrow-gap semiconductors is considered analogous to that of free-electron in vacuum predicted earlier by Kaplan. The resonance is based on the "Doppler" and "Lorentz" nonlinear mechanisms with a hysteresis due to the nonparabolicity of conduction band; it is excited by two laser beams when their frequency difference is near the cyclotron frequency.

In previous work[1] by one of the authors it was shown that due to relativistic effects, a large cyclotron motion of a free electron in vacuum can be excited by two laser beams with their frequencies much higher than a cyclotron frequency Ω. The laser frequencies must differ either by Ω or 2Ω which corresponds to either three-photon or four-photon resonance respectively. Although the multi-photon resonances of a single free electron can be caused by three different mechanisms (which were called Doppler, Lorentz and relativistic mechanisms respectively in Ref.[1]), the excited cyclotron motion always displays a hysteretic resonance based solely on the relativistic mass-effect. The hysteresis is a distinct feature of relativistic resonances regardless of whether they occur at the main frequency (i.e. when the driving frequency is near the cyclotron frequency) or at the combination frequency. The hysteretic resonance of a free slightly relativistic electron at the main frequency was early predicted by one of the authors[2] and subsequently observed experimentally[3]. It was also shown[4,2] that the hysteretic resonance of a similar nature at the main frequency may occur in narrow-gap semiconductors owing to the pseudorelativistic properties of their conduction electrons.

In this paper, we consider the feasibility of a hysteretic three-photon resonance in narrow-gap semiconductors analogous to that of free electron in vacuum[1]. The effect is feasible due to the nonparabolicity of the narrow-gap semiconductor conduction band which causes a pseudorelativistic dependence[5–7] of the effective mass of conduction electrons on their momentum or energy. The generation of a difference frequency $\Omega = \omega_1 - \omega_2$ using essentially the same mechanism of nonlinearity in a semiconductor by laser beams with frequencies ω_1 and ω_2 has already been demonstrated in the earlier work using either spin resonance[8], or cyclotron resonance[8–10]. However, in that earlier work[8–10], the hysteretic resonance has not been observed (or looked for) which might be attributed to the laser intensity apparently insufficient for such an effect. In this paper we show that this effect is feasible in narrow-gap semiconductors such as

InSb, GaAs, and HgTe, for example, under the pumping by two modes of CO_2 laser at $10.6\mu m$ and $9.4\mu m$ such that difference frequency corresponds to $\lambda \sim 83\mu m$, with the intensity of pumping being 10^5–$10^7 W/cm^2$.

We shall demonstrate the feasibility of this effect using a classical model for the interactions of optical waves with a single electron in the conduction band of the narrow-gap semiconductor. The thin semiconductor layer is immersed in the homogeneous magnetic field H_0. The semiconductor layer is also subject to the actions of two laser beams with frequencies ω_1 and ω_2 propagating normally to the field $\vec{H}_0$ (as in Ref.[1]).

Following the Kane two-band model[5] with isotropic bands, the energy of the conduction electrons in narrow-gap semiconductors can be expressed as

$$W(p) = (m^{*2}_0 v_0^4 + p^2 v_0^2)^{1/2}, \tag{1}$$

where $\vec{p}$ is the momentum of the conduction electron, m^*_0 is its effective mass at the bottom of the conduction band, $v_0 = (W_G/2m^*_0)^{1/2}$ is some characteristic speed, and W_G is the band gap [the energy W in Eq. (1) is measured with respect to the middle of the gap]. The velocity $\vec{v}$ of the conduction electron is given by[5] $\vec{v}(\vec{p}) = \partial W(p)/\partial \vec{p}$. By virtue of Eq. (1), this yields

$$\vec{p} = m^*_0 \vec{v}/(1 - v^2/v_0^2)^{1/2}\ , \vec{v} = \vec{p}/[m^*_0(1 + p^2/p_0^2)^{1/2}]\ , \tag{2}$$

where $p_0 = m^*_0 v_0 = (W_G m^*_0/2)^{1/2}$ is some characteristic momentum. One can see from Eqs.(1), (2) that relations among W, v, and p are completely relativistic, with v_0 playing a role as an "effective speed of light," and $W_G/2$ as "effective rest energy" of the electron. The motion of an electron in the semiconductor layer under the action of a homogeneous magnetic field $\vec{H}_0$ and an arbitrary number of plane EM waves is governed by the Lorentz equation with an additional damping term due to relaxation of momentum:

$$\frac{d\vec{p}}{dt} + \frac{1}{\tau}\,\vec{p} = e \sum_j \vec{E}_j + \frac{e}{c}\,\vec{v} \times (\vec{H}_0 + \sum_j \vec{H}_j)\,, \tag{3}$$

where τ is relaxation time of momentum which depends upon the scattering of electrons, e is the electron charge, $\vec{H}_0$ is a constant magnetic field, $\vec{E}_j$, $\vec{H}_j$ are respectively electric and magnetic fields of EM waves with their frequencies ω_j. For plane EM waves $\vec{H}_j = \sqrt{\epsilon_j}\,(\vec{k}_j/k_j \times \vec{E}_j)$ where $\epsilon_j = \epsilon(\omega_j)$ is the dielectric constant of semiconductor at the frequency ω_j. We introduce dimensionless momentum $\vec{\rho}$, electric fields $\vec{f}_j$, damping rate Γ, unity vector of magnetic field $\vec{h}$, unity wave vectors $\vec{q}_j$ and electron energy γ as follows:

$$\vec{\rho} = \frac{\vec{p}}{p_0}\ ;\ \vec{f}_j = \frac{e\vec{E}_j}{p_0 \Omega_0}\ ;\ \vec{h} = \frac{\vec{H}_0}{H_0}\ ;$$

$$\vec{q}_j = \frac{\vec{k}_j}{k_j}\ ;\ \gamma = (1 + \rho^2)^{1/2}\ ;\ \Gamma = (\tau\,\Omega_0)^{-1}\,, \tag{4}$$

where $\Omega_0 = eH_0/(m^*{}_0 c)$ is an unperturbated cyclotron frequency. The Lorentz equation (3) is then expressed in the form

$$\Omega_0^{-1}\,\dot{\vec{\rho}} + \Gamma\vec{\rho} = \sum_j \vec{f}_j +$$

$$+ (v_0/c)\,\gamma^{-1}\vec{\rho} \times \sum_j \sqrt{\epsilon_j}\,(\vec{q}_j \times \vec{f}_j) + \gamma^{-1}(\vec{\rho} \times \vec{h}) \tag{5}$$

which differs from the respective equation in Ref.[1] by factors v_0/c and $\sqrt{\epsilon_j}$, as well as by difference in defining ρ and f_j. It is worth noting that in narrow-band semiconductors $m^*_0 << m_0$ and $v_0 << c$ which implicates two important facts. Firstly, the cyclotron resonance can be observed for much higher frequencies (compared with free electrons) for the same magnetic field; for the magnetic fields currently available, it could be done on the infrared range. Secondly, the nonlinearity which is due to pseudorelativistic mass-effect scaled by c/v_0 becomes much larger than relativistic mass-effect of free electrons. This allows one to expect a reasonably low threshold of hysteretic excitation even considering a large increase of damping compared with free-electrons losses related to a synchrotron radiation.

We shall follow the approach developed in Ref.[1], whereupon the momentum $\vec{\rho}$ is assumed to be a sum of purely "cyclotron" component $\vec{\rho}_c$ with the frequency of rotation $\Omega = \Omega_0/\gamma_c$; where $\gamma_c = (1 + \rho_c^2)^{1/2}$, and all the other, noncyclotron components of various orders s, $\vec{\rho}_{nc}^{(s)}$. The motion $\vec{\rho}_c$ at the cyclotron frequency is determined then by the equation:

$$\Omega_0^{-1}(\frac{d\vec{\rho}_c}{dt}) - \gamma_c^{-1}[\vec{\rho}_c \times \vec{h}] + \Gamma\vec{\rho}_c = \vec{F}_c^{(1)}(t) + \vec{F}_c^{(2)}(t) + \cdots , \quad (6)$$

where the driving terms $\vec{F}^{(s)}$ in the right-hand part of this equation are nonlinear forces of different orders; "c" in $\vec{F}_c^{(s)}$ labels those components of these forces that oscillate with the cyclotron frequency Ω and are orthogonal to $\vec{H}_0$. In Eqs. (5) and (6) $\vec{\rho}_{nc}^{(s)}$ and $\vec{F}^{(s)}$ are defined as:

$$\vec{F}^{(1)} = \sum_{j} \vec{f}_j(\omega_j t - \vec{k}_j \cdot \vec{r}_c(t)) +$$

$$+ (v_0/c)\, \gamma_c^{-1}\, \vec{\rho}_c \times \sum_{j} \sqrt{\epsilon_j}\, (\vec{q}_j \times \vec{f}_j); \;\; (s = 1), \tag{7}$$

$$\Omega_0^{-1} \frac{d\vec{\rho}_{nc}^{(s)}}{dt} - \gamma_c^{-1}(\vec{\rho}_{nc}^{(s)} \times \vec{h}) + \gamma_c^{-3}(\vec{\rho}_{nc}^{(s)} \cdot \vec{\rho}_c)(\vec{\rho}_c \times \vec{h}) =$$

$$= \vec{F}^{(s)} - \vec{F}_c^{(s)}; \;\; (s \geq 1), \tag{8}$$

$$\vec{F}^{(s)} = \vec{F}_D^{(s)} + \vec{F}_L^{(s)} + \vec{F}_R^{(s)}; \;\; (s > 1)\,, \tag{9}$$

where

$$\vec{r}_c = \gamma_c^{-1}\, v_0 \int \vec{\rho}_c dt = -v_0(\Omega\gamma_c)^{-1}[\vec{\rho}_c \times \vec{h}]\,, \tag{10}$$

and forces $\vec{F}_D$, $\vec{F}_L$, $\vec{F}_R$ are originated by the first, second, and third terms in right-hand side of Eq.(5), and their nature can be regarded[1] as the "Doppler", "Lorentz", and "relativistic" mechanisms respectively. Which one of these particular mechanisms dominates, depends on the fields propagation configuration and their polarizations. In particular, the force $\vec{F}^{(2)}$ is responsible for for the three-photon and four-photon resonances[1], is determined as follows:

$$\vec{F}^{(2)} = -\frac{v_0}{c}\, \gamma_c^{-1} \sum_{j} \sqrt{\epsilon_j}\, (\vec{q}_j \cdot \int \vec{\rho}_{nc}^{(1)} dt)\, \frac{\partial \vec{f}_j}{\partial t}$$

$$+ \frac{v_0}{c}\, \gamma_c^{-1}\, \vec{\rho}_{nc}^{(1)} \times \sum_{j} \sqrt{\epsilon_j}(\vec{q}_j \times \vec{f}_j) -$$

$$- \gamma_c^{-3} \left\{ \vec{\rho}_{nc}^{(1)}(\vec{\rho}_c \cdot \vec{\rho}_{nc}^{(1)}) + \vec{\rho}_c/2\, [(\vec{\rho}_{nc}^{(1)})^2 - 3\gamma_c^{-2}\, (\vec{\rho}_c \cdot \vec{\rho}_{nc}^{(1)})^2] \right\} \times \vec{h}\,. \tag{11}$$

In order to find a threshold of hysteretic three-photon resonance and to describe a resonance with comparatively low energy and momentum [$\rho_c << c\Omega_0/(v_0\omega_j\sqrt{\epsilon_j})$ which amounts to $r_c << k_j^{-1}$], we can neglect the third term in a right-hand side at Eq.(11) and assume $\gamma_c \approx 1$ which results in

$$\vec{F}^{(2)} = -\frac{v_0}{c}\sum_j \sqrt{\epsilon_j}\left[(\vec{q}_j \cdot \int \vec{\rho}_{nc}^{(1)} dt)\,\frac{\partial \vec{f}_j}{\partial t} - \vec{q}_j\,(\vec{f}_j \cdot \vec{\rho}_{nc}^{(1)})\right]. \quad (12)$$

In this approximation, only Doppler and Lorentz forces are responsible for the nonlinear excitation; the contributions from both of them are of the same order of magnitude. It is worth emphasizing, though, that once the cyclotron motion is excited, it is only the relativistic-mass effect [the term γ_c^{-1} on the left-hand side of Eq.(6)] that acts to limit the energy of excitation and to form a hysteretic resonance.

We consider now three-photon excitation ($\Omega = \omega_1 - \omega_2$) for the specific configuration when two waves with their respective frequencies ω_1 and ω_2 counter-propagate normally to the magnetic field $\vec{H}_0$ (e.g. along the axis x, i.e. $\vec{q}_1 = -\vec{q}_2 = \vec{e}_x$) with their polarizations parallel to each other and normal to $\vec{H}_0$, such that $\vec{f}_{1,2} = \vec{e}_y\, f_{1,2} \sin(\omega_{1,2}t \mp k_{1,2}x_c)$. Neglecting second term in the right-hand side of Eq.(7), (since $v_0/c << 1$), we obtain $\vec{F}^{(1)} \approx (f_1 \sin\omega_1 t + f_2 \sin\omega_2 t)\,\vec{e}_y$. Solving now Eq.(8) for $\vec{\rho}_{nc}^{(1)}$ in the approximation $r_c << k_{1,2}^{-1}$ as

$$\vec{\rho}_{nc}^{(1)} \approx -[(\Omega_0 f_1/\omega_1)\cos\omega_1 t + (\Omega_0 f_2/\omega_2)\cos\omega_2 t]\,\vec{e}_y \quad (13)$$

and inserting it into Eq.(12), we obtain the required nonlinear force $F_c^{(2)}$ as

$$\vec{F}_c^{(2)} = \frac{f_1 f_2 v_0 \Omega_0}{2c} \left(\sqrt{\epsilon_1}/\omega_1 + \sqrt{\epsilon_2}/\omega_2\right) \sin\Omega t \, \vec{e}_x \tag{14}$$

Assuming now in Eq.(6) that $\vec{\rho}_c = \rho_c[\vec{e}_x \sin(\Omega t + \phi) + \vec{e}_y \cos(\Omega t + \phi)]$, we arrive at the dynamic equations for amplitude ρ_c and phase ϕ of the cyclotron excitation:

$$\Omega_0^{-1} \frac{d\rho_c}{dt} = -\Gamma\rho_c + \mu_{(3)} \cos\phi \tag{15}$$

$$-\Omega_0^{-1} \frac{d\phi}{dt} = (\Omega - \Omega_0)/\Omega_0 + \rho_c^2/2 + \mu_{(3)} (\sin\phi)/\rho_c \tag{16}$$

where we introduced a three-photon driving parameter $\mu_{(3)}$:

$$\mu_{(3)} = \frac{1}{4} f_1 f_2 \, \Omega_0 \, \frac{v_0}{c} \, [\sqrt{\epsilon_1}/\omega_1 + \sqrt{\epsilon_2}/\omega_2] \, . \tag{17}$$

The stationary regime is achieved when $d/dt = 0$; the steady-state amplitude ρ_c is determined then by the equation

$$\rho_c = \mu_{(3)} \, [\Gamma^2 + (\Omega/\Omega_0 - 1 + \rho_c^2/2)^2]^{-1/2} \tag{18}$$

analogous to Eq.(8) of Ref.[1]. Under appropriate conditions, this equation has three solutions (two of them stable, and one - unstable) which results in hysteretic behavior of stationary excitation and therefore, in bistability. In order to obtain a hysteresis, one has to have $\mu_{(3)}$ exceed a threshold

$$\mu_{cr} \approx 1.75 \, \Gamma^{3/2} \tag{19}$$

To estimate the critical laser intensity implied by this condition, we consider three different semiconductors (InSb, GaAs, and HgTe) pumped by two different lines at CO_2 laser at wavelengths $\lambda_1 = 10.6\mu m$ and $9.1\mu m$ with the same intensity of both of them (i.e.

$f_1 = f_2$), such that the difference frequency of cyclotron excitation is located in a very far infrared range, $\lambda_0 = (\lambda_2^{-1} - \lambda_1^{-1})^{-1} = 83.03\mu m$. The magnetic field required for this cyclotron frequency, depends on the effective mass of conduction electron, m^*_0, and is different for each of those semiconductors. Therefore, the magnetic field H_0 resulting in the cyclotron wavelength $\lambda_0 = 83.03\mu m$ is 1.95 for InSb, 8.51 for GaAs, and 3.74 Tesla for HgTe, respectively. The required temperature as well as data on energy gaps and dielectric constants can be found in Table 1 (we assume that dielectric constant is the same at the both laser frequencies ω_1 and ω_2 since they are pretty close).

Table 1.

Mater.	Temp. Need (K)	Effect. Mass Ratio (m^*_0/m_0)	Energy Gap (eV)	Dielec. Const.	Thresh. Laser Inten. (W/cm^2)
InSb	4.2	0.015[12]	0.237[12]	15.7[11]	4.3×10^4
GaAs	77	0.066[11]	1.54[11]	10.9[11]	1.63×10^6
HgTe	10	0.029[13]	0.283[12]	19[11]	1.54×10^5

For all the three cases, when evaluating damping parameter $\Gamma = 1/\tau\Omega_0$, we assume relaxation time of momentum τ to be roughly the same, $\tau \approx 10^{-11}$ sec, which was directly measured for InSb[12], and can be obtained from the data on the Hall mobility μ for GaAs[11] and HgTe[13] as $\tau = m^*_0\mu/e$. When evaluating the critical intensity of the laser field through Eqs.(17) and (19), using definition for f, Eq.(4), we notice that the microscopic fields E_j in Eqs.(3) and (4) are related to the macroscopic laser field E_L by the Clausius-Mossotti formula $E_j = E_L(\epsilon_j + 2)/3$. The resulting critical intensity of laser required to obtain a hysteresis is shown in the last column in Table 1.

As far as possible experiment on observation of hysteresis is concerned, the important question is whether other nonlinear effects (first of all, the *interband* multi-photon processes in the case of InSb) will provide a strong competition to the nonlinear intraband processes in consideration (which might be regarded as multi-photon *intraband* process). However the analysis of the experiment[9] and data[10] clearly shows that in the required range of pumping laser intensities, the interband two-photon and three-photon absorption could be neglected. Indeed, one may see from the Table 1 that for InSb, the critical intensity required for hysteresis (provided that the relaxation time is $\sim 10^{-11}$s as in Ref.[12]) is 4.3×10^4W/cm^2. This is less than 10^5W/cm^2 used in experiment[9] which explored exactly the same mechanism of nonlinearity that is being proposed by us to use for observation of hysteresis in this paper[14]. Therefore, in the range of intensities $\lesssim 10^5$W/cm^2 the intraband nonparabolic effects in InSb can be expected to dominate the interband multi-photon effects which makes it feasible to observe pseudorelativistic- related hysteresis for appropriately chosen samples of InSb.

Another material, HgTe, at a first glance might seem inappropriate for observation of the nonparabolicity-related hysteresis since it is known as zero-gap material with a negative energy gap[15]. The interesting fact though is that nonparabolicity of the conduction band in HgTe can still be approximated by Kane's model, whereby W_G in Eqs.(1) and (2) (for $E/W_G << 1$) should be interpreted[15] now as the minimum energy difference between valence bands Γ_6 and Γ_8. Indeed, it was shown empirically[15] that if W_G is interpreted this way, the effective mass of the conduction electron can be expressed as $m^* \approx m^*_0(1 + 2E/W_G)$ where E is the kinetic energy of the conduction electron which exactly comes to Eqs.(1) and (2) resulting from Kane's model. In the experiment, however, the two-photon transition between valence bands may result in the increased absorption of the pumping laser radiation which might *not* be prohibiting for the hysteresis, since the relaxation time (and dipole momentum) of the interband two-photon transition and those of intraband processes are very different. This question must be addressed in the future calculations for the specific conditions of experiment with HgTe.

The attractive feature of GaAs is relatively high temperature ($\sim$ 77K, see Table 1) sufficient for experiment. In addition, GaAs does not suffer from the interband multi-photon processes since the energy gap is quite large compared with 0.1eV corresponding to the CO_2 laser energy. The nonparabolicity is consequently small and therefore the laser threshold intensity is high compared with other materials (see Table 1). Notwithstanding this fact, the considerable excitation-related change of effective mass of conduction electron has been observed in the most recent experiment[16]. A feasible perturbing mechanism in GaAs, which is a polar crystal, is the excitation of optical phonons by laser radiation. The recent experimental research[17] showed, however, that the optical phonons in GaAs can not be excited even at $\lambda \sim 1\mu m$ which implicates their nonexistence for longer wavelength, in particular, for CO_2 laser.

When the driving parameter $\mu_{(3)}$ considerably exceeds the threshold μ_{cr}, the kinetic energy of the cyclotron motion, $\rho_c^2/2$, follows almost exactly the resonant detuning $\rho_c^2/2 = (\Omega_0 - \Omega)/\Omega_0$ (for $\Omega < \Omega_0$), until it reaches its maximum magnitude, $\rho_{c\max}^2 = (\mu_{(3)}/\Gamma)^2$, which happens at $(\Omega_0 - \Omega)/\Omega_0 = (\mu_{(3)}/\Gamma)\sqrt{2}$. Immediately after that, if $|\Omega - \Omega_0|$ continues to increase, the electron jumps from the higher excitation branch down to almost zero excitation. A small-perturbation analysis of Eqs.(15) and (16) reveals that the third branch of the steady-state solution, Eq.(18), located between the higher and lower branches (both of which are stable) is unstable.

In conclusion, we demonstrated the feasibility to observe a hysteretic (bistable) three-photon resonance in narrow-gap semiconductors using two laser beams with their difference frequency near the cyclotron frequency. For example, using two lines of a CO_2 laser, a large hysteretic excitation may be observed in a very far infrared range with the laser intensity of 10^5–$10^7 W/cm^2$. This effect may be of significant interest both for the study of highly excited Landau levels and for such applications as optical bistability and far-infrared sources of coherent radiation.

This work is supported by the U.S. Air Force Office of Scientific Research.

References

[1] A. E. Kaplan, Phys. Rev. Lett. 56, 456 (1986).

[2] A. E. Kaplan, Phys. Rev. Lett. 48, 138 (1982); A. E. Kaplan, IEEE J. Quant. Electr. QE–21, 1544 (1985); A. E. Kaplan, Nature, 317, 476 (1985).

[3] G. Gabrielse, H. Dehmelt, and W. Kells, Phys. Rev. Lett. 54, 537 (1985).

[4] A. E. Kaplan and A. Elci, Phys. Rev. B29, 820 (1984).

[5] E. O. Kane, J. Phys. Chem. Solids 1, 249 (1957).

[6] B. Lax, in *Proceedings of the Seventh International Conference on the Physics of Semiconductors, Paris, 1964* (Dunod, Paris, 1964), p. 253; A.G. Aronov, Fiz. Tverd. Tela (Leningrad) 5, 552 (1963) [Sov. Phys. — Solid State 5, 402 (1963)]; H.C. Proddaude, Phys. Rev. A140, 1292 (1965); W. Zawadzki and B. Lax, Phys. Rev. Lett. 16, 1001 (1966).

[7] W. Zawadzki, S. Klahn, and U. Merkt, Phys. Rev. Lett. 55, 983 (1985).

[8] P. A. Wolff, IEEE J. Quant. Electr., QE–2, 659 (1966); V. T. Nguyen and T. J. Bridges, Phys. Rev. Lett. 29, 359 (1972); T. L. Brown and P. A. Wolff, Phys. Rev. Lett. 29, 362 (1972).

[9] E. Yablonovitch, N. Bloembergen, and J. J. Wynne, Phys. Rev. B3, 2060 (1971).

[10] J. J. Wynne, Phys. Rev. B6, 534 (1972).

[11] D. L. Rode, *Semiconductors and Semimetals,* edited by R. K. Willardson and A. C. Beer, Vol. 10, *Transport Phenomena,* Academic Press, New York (1975).

[12] O. Matsuda and E. Otsuka, J. Phys. Chem. Solids 40, 815 (1979).

[13] R. A. Stradling and G. A. Antcliffe, J. Phys. Soc. Jap. Suppl. 21, 374 (1966).

[14] The possible reason why the hysteresis has not been observed in the experiment[9] is that the samples used in [9] had much higher carrier concentration as well as high temperature compared with Ref.[12] used by us for estimation based on Eq.(19). These differences result in critical intensity required to observe a hysteresis under conditions used in Ref.[9], much higher than that one used in Ref.[9].

[15] J. Tuchendler, M. Grynberg, Y. Couder, H. Thome, and R. L. Toullec, Phys. Rev. B8, 3884 (1973); P. Kacman and W. Zawadzki, Phys. Status Solidi 47, 629 (1971); Y. Guldner, C. Rigaux, M. Grynberg, and A. Mycielski, Phys. Rev. B8, 3875 (1973).

[16] M. Heiblum, M. V. Fischetti, W. P. Dumke, D. J. Frank, I. M. Andersen, C. M. Knoedler, and L. Osterling, Phys. Rev. Lett. 58, 816 (1987).

[17] C. L. Collins and P. Y. Yu, Phys. Rev. B30, 4501 (1984); D. Von der Linde, J. Kuhl, and H. Klingenburg, Phys. Rev. Lett. 44, 1505 (1980); K. T. Tsen, D. A. Abramsohn, and R. Bray, Phys. Rev. B26, 4770 (1982); J. Shah, R. C. C. Leite, and J. F. Scott, Solid State Commun. 8, 1089 (1970).

The Nonlinear Susceptibility due to Interband Transition in Semiconductors and the Optical Bistability

Ou Fa Zhang Xiao-dong Wu Ting-wan

Physics Department
South China Institute of Technology
Guangzhou, PRC

The nonlinear susceptibility due to interband transition in semiconductors is calculated in this paper. According to this susceptibility, we have studied the optical bistability of InSb. The theoretical bistable curve and the theoretical relation between the critical switching intensity and the wavenumber of incident light are consistent with the published results.

I. Introduction

The nonlinear susceptibility due to interband transition in semiconductors was ever calculated by B.S.Wherret and N.A.Higgins based on the two-band model[1]. We found that the real part of the susceptibility χ_r in that paper is actually divergent. We have improved the two-band model and renewed the calculation of the susceptibility. A more reasonable convergent results are obtained. According to our obtained susceptibility, we deal with the optical bistability due to band-filling in semiconductors analytically. The theoretical results agree with the published experimental results for InSb[2][3].

II. The Improved Two-band Model and the Susceptibility due to Interband Direct Transition in Semiconductors

The interband direct transition means the transition between the two levels that belong to the valence band and the conduction band respectively and that have the same quasi-momentum $\hbar\vec{k}$. The energy interval between this pair of levels is $E(\vec{k}) = E_c(\vec{k}) - E_v(\vec{k})$. Expanding $E(\vec{k})$ at $\vec{k}_o$, at which $\nabla_{\vec{k}} E(\vec{k}) = 0$, neglecting $(\vec{k} - \vec{k}_o)^3$ and its higher order terms, and assuming the effective mass of electron or hole (m_e or m_h) to be isotropy, then $E(\vec{k})$ can be written as

$$E(\vec{k}) = E(\vec{k}_o) + \frac{\hbar^2(\vec{k} - \vec{k}_o)^2}{2m_r} , \qquad (1)$$

where m_r is the reduced mass, $E(\vec{k}_o)$ is the gap width Eg :

$$m_r^{-1} = m_e^{-1} + m_h^{-1} , \qquad E(\vec{k}_o) = Eg . \qquad (2)$$

It is necessary to point out that the equation (1) can only be used strictly to the Brillouin Zone near the bottom of the conduction band or the top of the valence band. However, in Reference[1], the wavevector $\vec{k}$ in equation(1)was extended to infinite, and it means that the bandwidth is infinite. In our opinion, this approach is far from the practice and leaves the divergent origin for their results.

Nevertheless, in order to simplify the calculation, the adequate extension for the $\vec{k}$ in equation (1) would be allowed. We intend to extend $\vec{k}$ to the top of the conduction band and the bottom of the valence band, which is shown in Fig.1 *) with broken curves, whereas the solid curves represent the actual band structure. Certainly, this extension could bring about some errors. At last, we assume that the states above the top of conduction band are ionized states with the continuous spectrum, then the conduction band width can be confined by the affinity energy of electron E_a. Without lossing generality, let $\vec{k}_o = 0$, under above conditions,

$$E_a = \frac{\hbar^2}{2m_e} k_{max}^2 , \qquad (k_{max} = |\vec{k}_{max}|) \qquad (3)$$

where the meaning of $\vec{k}_{max}$ has been shown in Fig. 1 .

Above discription is the statement of our improved two-band model for band-filling type semiconductors. According to the model in this

*) All the figures are collected in the last of this paper.

paper, we have calculated the nonlinear susceptibility due to inter-band direct transition. The selection rule of interband direct transition, $\vec{k}_c = \vec{k}_v = \vec{k}$, means that the two bands (conduction and valence band) can be taken as a set of the inhomogeneously broadened two-level system. Thus, the susceptibility due to direct transition in semiconductors can be expressed as

$$\chi(\omega) = \int_{\omega_g}^{\omega_{o\,max}} \chi(\omega,\omega_o) g(\omega_o)\, d\omega_o \quad , \tag{4}$$

where ω is the frequency of incident light. According to the equation (1),(2),(3),

$$\omega_o = \frac{E(k)}{\hbar} = \omega_g + \frac{\hbar}{2m_r} k^2 , \qquad \omega_g = \frac{E_g}{\hbar} \tag{5}$$

$$\omega_{o\,max} = \omega_g + \omega_a\left(1 + \frac{m_e}{m_h} \right) , \qquad \omega_a = \frac{E_a}{\hbar} \tag{6}$$

$\chi(\omega,\omega_o)$ is the susceptibility contributed by one pair of levels which are characterized by $\vec{k}$, $g(\omega_o)$ is the inhomogeneously brodened line shape function, and $g(\omega_o)d\omega_o$ represents the number of $\vec{k}$ in range ω_o —— $\omega_o + d\omega_o$ per unit volume. Considering that the isoenergetic surface is assumed to be sphere in $\vec{k}$ space, it is easy to get

$$g(\omega_o)d\omega_o = \frac{k^2}{\pi^2}\, dk \qquad (k = |\vec{k}|) . \tag{7}$$

According to the Reference[4],

$$\chi(\omega,\omega_o) = \frac{\mu^2 T_2^2}{\hbar} \cdot \frac{1}{\omega_o - \omega - iT_2^{-1}} \cdot \frac{(\omega_o-\omega)^2 + T_2^2}{1 + (\omega_o-\omega)^2 T_2^2 + T_1 T_2 \mu^2 \varepsilon^2/\hbar^2} , \tag{8}$$

where T_1 and T_2 are longitudinal and transverse relaxation time of the system, respectively, μ is the element of transition matrix of two-level atoms, ε is the amplitude of light field in the medium. Substituting equation (8) into (4) , using relation (3), (5), (6), (7), replacing the ω_o in equation (4) by k , then we divided the $\chi(\omega)$ into real part χ_r and imaginary part χ_i , the nonlinear susceptibility due to direct transition in semiconductors can be writen as following analytical expressions:

$$\chi_r(\mathcal{I}) = \frac{\sqrt{2\tilde{k}}}{4\pi}\left(\frac{2m_r}{\hbar T_2}\right)^{3/2}\left(\frac{2\sqrt{2}\, X_{max}}{\pi} - \left(\Delta + \sqrt{(1+\Delta^2+\mathcal{I})}\right)^{\frac{1}{2}}\right) \tag{9a}$$

$$\chi_i(\mathcal{I}) = \frac{\sqrt{2\tilde{k}}}{4\pi}\left(\frac{2m_r}{\hbar T_2}\right)^{3/2}\left(\sqrt{1+\Delta^2+\mathcal{I}} - \Delta\right)^{3/2} \Big/ \sqrt{1+\mathcal{I}} \tag{9b}$$

where

$$\mathcal{I} = T_1 T_2 \mu^2 \mathcal{E}^2/\hbar^2 = \text{Normalized intensity of light} \tag{10}$$

$$\Delta = (\omega_g - \omega) T_2 = \text{Normalized gap detuning} \tag{11}$$

$$X_{max} = \sqrt{\frac{\hbar T_2}{2m_r}}\, k_{max}, \quad \tilde{k} = \frac{\mu^2 T_2}{\hbar} \,. \tag{12}$$

III. Optical Bistability

1. Steady-State Properties

Let us study the dispersive OB in consideration of the linear absorption in F-P cavity based on the expression (9a) for $\chi_r(\mathcal{I})$. In the case of F-P cavity the relation between the incident light intensity I_i and transmitted I_t is [5]:

$$I_i = aI_t + FI_t \sin^2 \frac{2\pi n l}{\lambda} \tag{13}$$

where

$$a = \frac{(1-R(1-G))^2}{(1-G)(1-R)}, \quad G = 1-e^{-\alpha l}, \quad F = \frac{4R}{(1-R)^2}, \tag{14}$$

α -- coefficient of linear absorption, l -- length of the sample, R -- reflectivity of the cavity, λ -- wave length of the incident light and n -- nonlinear refrative index of the sample. Denoting n_b for background refrative index, we have

$$n(\mathcal{I}) \doteq n_b + \frac{2\pi \chi_r(\mathcal{I})}{n_b} \tag{15}$$

where $\chi_r(\mathcal{I})$ is determined by expression (9a). Combining eq.(13) with (15) and (9a), finally we get

$$I_i = aI_t + FI_t \sin^2\left\{\frac{2\pi l}{\lambda}\left[n_b + \frac{2\pi A}{n_b}\left(B - \left(\Delta + \sqrt{1+\Delta^2+\frac{I_t(1+R)}{I_S(1-R)}}\right)^{\frac{1}{2}}\right)\right]\right\} \tag{16}$$

where we have used the relation between the average light intensity in F-P cavity I_c and transmitted I_t [2],

$$I_c = \frac{1+R}{1-R} I_t, \text{ and } \mathcal{I} = \frac{I_c}{I_S} \tag{17}$$

where I_S — a constant as the saturated intensity. For A and B in (16), they are $A = \frac{\sqrt{2\tilde{k}}}{4\pi}\left(\frac{2m_r}{\hbar T_2}\right)^{3/2}$, $B = \frac{2\sqrt{2}}{\pi} X_{max}$.

The stationary equation (16) will reveal the OB due to band-filling in semiconductors.

It is necessary to examine the reliability of eq.(16). To this end in the benefit of following data for semiconductor InSb with narrow band gap and data for F-P etalon: $n_b \doteq 4$, $I_s = 1.5\times10^{-20}/T_1 T_2\,(w.cm^{-2})$, $\tilde{K} = 3\times10^{-5}T_2(cm^3)$, $T_2 = 10^{-12}s$ in 5^oK [1], $G = 1-e^{-\alpha l} = 0.5$—$0.6$ with $l = 560\mu m$, $\omega_g/c = 1899cm^{-1}$, $F = 0.5$ with $R \doteq 0.14$, $2\pi/\lambda = 1895cm^{-1}$ [2]; $m_e = 0.015m_o$, $m_h = 0.5m_o$ [6]; $E_a = 4.7ev$ [7], $T_1 = 10^{-10}s$ [8]; according to the eq.(16), the I_t vs I_i curve is drawn by computer, as given in Fig.2. It shows that our results have a good consistency with the experimental curve which is shown in Fig.3. Furthermore, it seems that in the respect of consistency with practice (at least for InSb) our theoretical multi-OB curve is better than that of ref.[9](see Fig.4) which is obtained by the fully quantum approach.

2. Critical switching irradiance

For the latter discussion, the generalized dynamic equation of OB under the conditions of mean field approximation and good-quality cavity is given as [10]

$$\frac{dx}{dt} = -k_c((1+2D\chi_i)x - y\cos\phi) \tag{18a}$$

$$\frac{d\phi}{dt} = k_c(2D\chi_r - \theta - \frac{y}{x}\sin\phi) \tag{18b}$$

where y, x and ϕ, θ are the normolized incident field amplitude, transmitted field amplitude and phase, cavety detuning respectively; k_c — decay rate of the cavity; D — $\omega l/4cT$, $T=1-R$. Putting $dx/dt = d\phi/dt = \chi_i = 0$ and substituting the expression (9a) into (18a), we obtain the dispersive OB stationary equation for the band-filling type semiconductors

$$Y = X\left\{1+\left[\theta - 2DA(B-(\Delta+\sqrt{\Delta^2+1+TX})^{\frac{1}{2}})\right]^2\right\} \tag{19}$$

where $Y = y^2$, $X = x^2$, $TX = Ic/Is$. It is assumed that

$$\Delta^2 >> 1+TX, \text{ and } \Delta^{3\frac{1}{2}} >> 2DA\ . \tag{20}$$

The above relations could be satistied in the optical region, because Δ^2 is about 10^4—10^6 in this region. Under these conditions, eq.(19) may be reduced approximatly as following:

$$Y = X\left\{1+\tilde{K}^2+\frac{\tilde{K}C\sqrt{2\Delta}}{2\Delta^2}+\frac{C^2}{8\Delta^3}+\left(\frac{\tilde{K}C\sqrt{2\Delta}T}{2\Delta^2}+\frac{C^2T}{4\Delta^3}\right)X+\frac{C^2T^2}{8\Delta^3}X^2\right\}, \tag{21}$$

where $C = DA$ is defined as the cooperative parameter, i.e.:

$$C = \frac{\omega l A}{4cT}\ , \text{ and } \widetilde{K} = \theta - 2CB + 2C\sqrt{2\Delta}\ . \tag{22}$$

From eq.(21), it is easy to get the OB condition[11]:

$$\widetilde{K} + \frac{C}{2\sqrt{2}\,\Delta^{3/2}} < -\sqrt{3}\ . \tag{23}$$

Based on the bistable condition, we can discuss the relation between the critical switching irradance and the frequency of incident light in the optical bistable system due to band-filling in semiconductors. According to the defination in Reference [3], critical switching irradance is defined as the input intensity that leads just to the appearance of the bistability. Let

$$\widetilde{K} + \frac{C}{2\sqrt{2}\,\Delta^{3/2}} = -\sqrt{3}\ ,$$

we obtain

$$X_c = \frac{8\sqrt{3}\,\Delta^{3/2}}{3\sqrt{2}\,CT}\ . \tag{24}$$

Substituting eq.(24) into (21), then

$$Y_c = \frac{32\sqrt{3}\,\Delta^{3/2}}{9\sqrt{2}\,CT} = \frac{16\sqrt{2}(\omega_g-\omega)^{3/2}\,T_2^{3/2}}{3\sqrt{3}\,CT}\ , \tag{25}$$

where Y_c is the normalized critical switching irradiance, T is the transmittance of the mirrors. For InSb, Y_c — k theoretical curve consists with the I_c — k experimental curve qualitatively[3] (see Fig.5 & Fig.6, where k is the wavenumber of incident light). This result indicates that the equation derived in the literature [10] is correct. If we take InSb as Kerr medium ($n=n_0+n_2 I$, n_2 is given by experiment), then we can not theoretically discuss the relation between the critical switching irradiance and the frequency of incident light. Thus, it has certain advantage to use the susceptibility to deal with the optical bistability due to band-filling in semiconductors.

3. Transient properties (based on the dynamical eqs.(18a,b))

With numerical simulation for InSb, the following transient properties (Cf., for example, Fig.7 and Fig.8 for jumping up process) in the band-filling type optical bistable system have been found:

1) Critical slowing down

For jumping up process, the closer to the upper threshold, the ultimate value of incident light is, the longer the switching time is. For jumping down process, the closer to the **lower threshold** the ultimate value of incident light is, the longer the switching time is.

2) Overstate phenomena

Some transient values in the jumping up process are larger than the upper stable value, some transient values in the jumping down process are smaller than the lower stable value.

3) Relation between the initial value of incident light and switching time

If the initial value of incident light is smaller, the jumping up time will be shorter. If the initial value of incident light is larger, i.e. closer to the upper threshold, the jumping up time is longer. When the initial value of incident light is far from the lower threshold, the jumping down time is shorter, when the initial value of incident light is closer to the lower threshold, the jumping down time is longer.

References

[1] B.S.Wherrett, N.A.Higgins, Proc. R.Soc.Lond A. 379, 67, (1982)

[2] D.A.B.Miller, et al., Appl. Phys. Lett. Vol.35, pp658-60 (1979)

[3] H.A.Alattar, et al., Optics Commum. Vol.58,NO.6 (1986)

[4] A.Yariv, Quantum Electronics, Second edition,New York 1975

[5] A.Yariv, Optical Electronics, Second edition, N.Y. 1976

[6] Ye Shi Zhong, " Compound Semiconductor Material and its Application", Machine Industry Press (1986) , published in chinese.

[7] Liu En Ke, "Semiconductor Physics", National Defence Industry Press (1979) , published in chinese.

[8] D.A.B,Miller, et al., Optical Bistability, ed. by C.M.Bowden, p115 (1980)

[9] J.Gol, H.Haken, Phys. Rev. A, vol.28, No.2. 910 (1983)

[10] Ou Fa, et al., to be published

[11] G.P.Agrawal, et al., Phys. Rev. A, 19 5 2074-85 (1979)

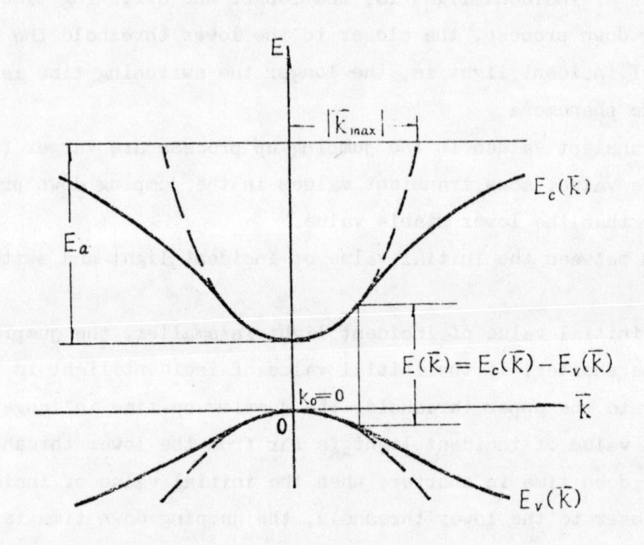

Fig.1 The Improved Two-band Model

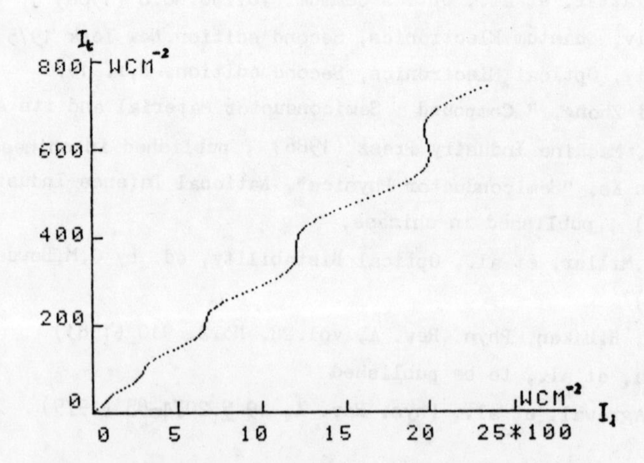

Fig.2 Theoretical curve of bistability for InSb at 5k.

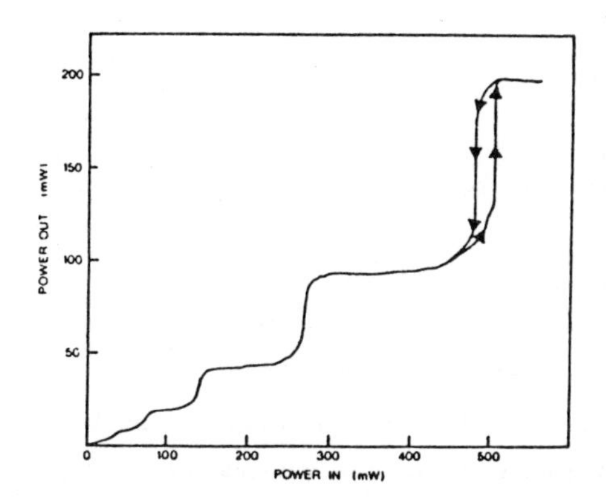

Fig.3 Experimental curve of bistability for InSb at 5k, from Reference[2].

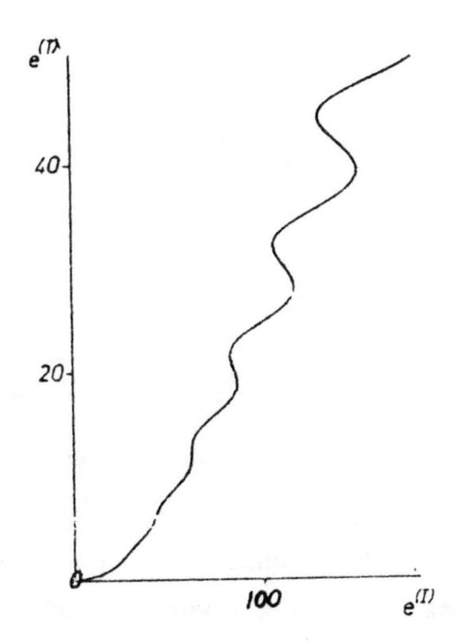

Fig.4 Theoretical optical bistable curve obtained by J.Goll and H.Haken (1983).

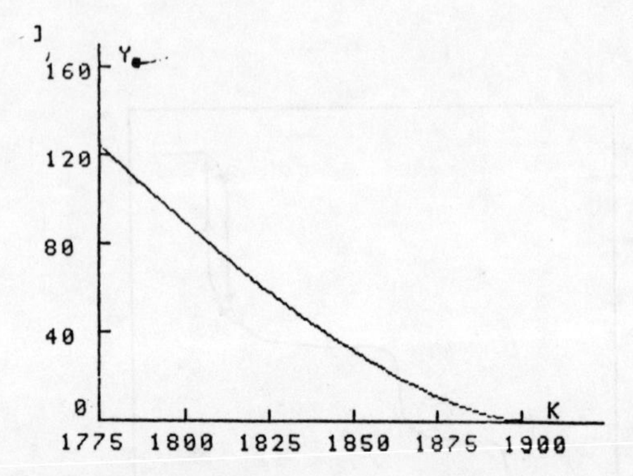

Fig.5 Relation between the critical switching irradiance and the wavenumber of incident light for InSb.

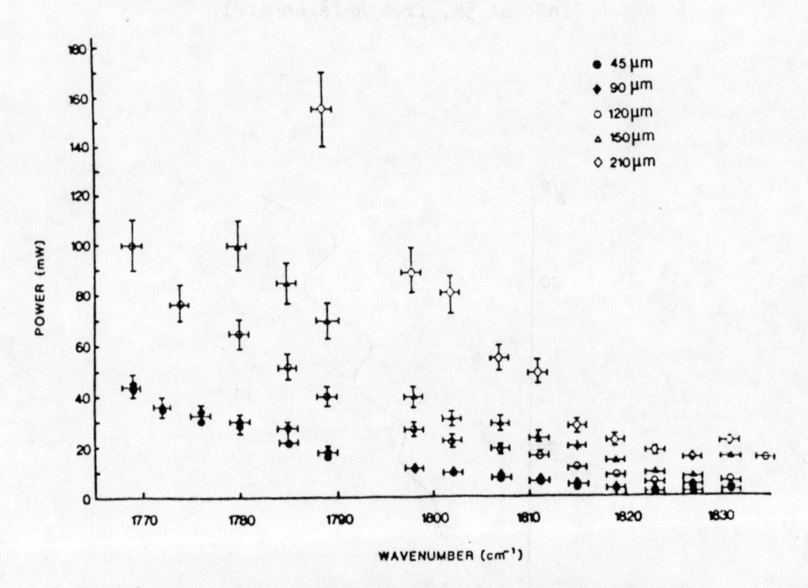

Fig.6 Frequency dependence of the experimental critical switching power for InSb. The different curves are obtained for different light spot sizes, from Reference[3].

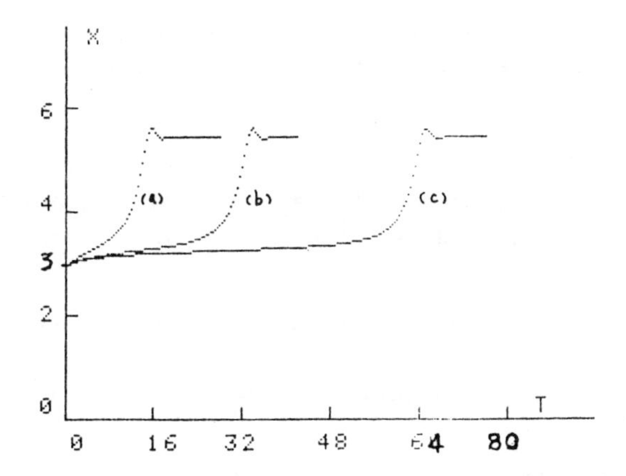

Fig.7 Time evolution of the transmitted field for $y_o = 5.54$, $y'_{\downarrow} = 5.92$; (a) $y' = 5.982$ (b) $y' = 5.935$ (c) $y' = 5.925$. Time is in units of k^{-1}.

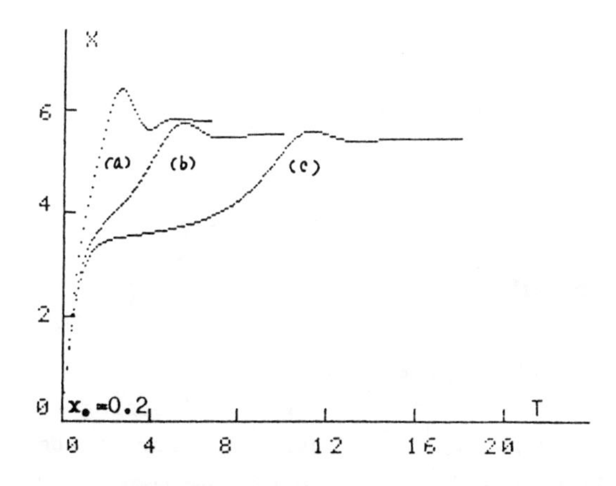

Fig.8 Time evolution of the transmitted field for $y_o = 0.4$, $y'_{\uparrow} = 5.92$; (a) $y' = 7.625$ (b) $y' = 6.225$ (c) $y' = 5.925$. Time is in units of k^{-1}.

Analysis of Optically Bistable Systems Using Ring Cavity With Two Partially Reflecting Mirrors of Unequal Transmission Coefficients

Chen Jishu[1], Luo Liguo[2]

[1] Physics Department, Ningbo University, Ningbo Zhejiang, China
[2] Optics Department, Shandong University, Jinan Shandong, China

When the transmission coefficients T1 T2 of the two partially reflecting mirrors in an optically bistable system with ring cavity filled by homogeneously broadened two-level atomic medium are such that

$$0 < T1/T2 \equiv N < \infty \quad \text{(together with } T3 = T4 = 0\text{)}$$

the steady state equation in the purely absorption and resonant cases can be shown to be

$$y = x_T\sqrt{N}\left[\frac{1}{2}\left(1+\frac{1}{N}\right) + \frac{2C}{1+N x_T^2}\right] \tag{1}$$

$$y = \frac{1}{2}\left(1+\frac{1}{N}\right)(x_R + y) + \frac{2C(x_R+y)}{1+(x_R+y)^2} \tag{2}$$

in which all notations are as usual, e.g., y denotes input field, x_T transmission output field and x_R reflecting output field. (see Fig. 1 and Fig.2)

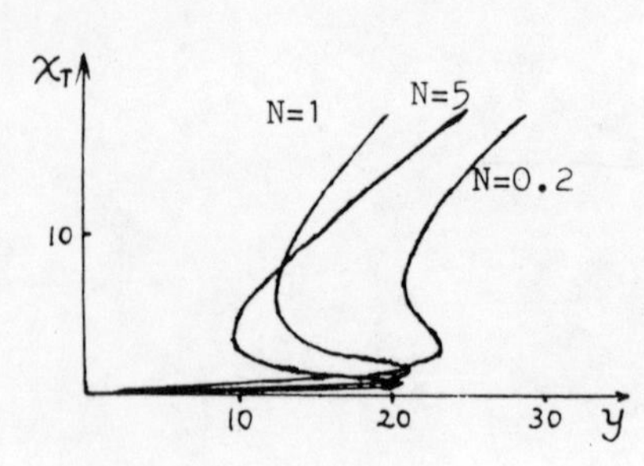

Fig.1. Plot of output versus input light amplitudes at steady state for purely absorptive cases (C=20).

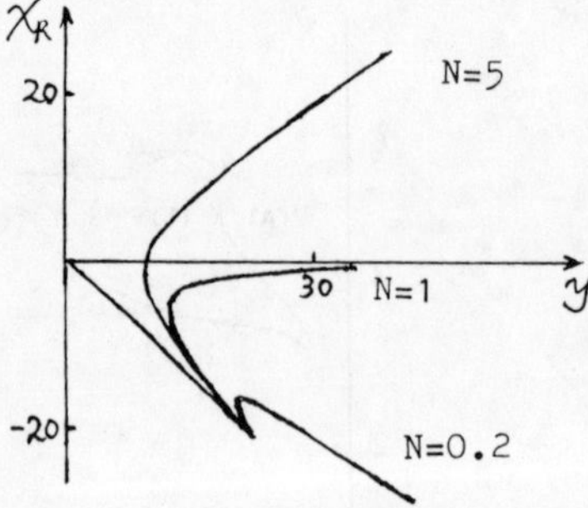

Fig.2. Reflecting steady state curve for purely absorptive cases (C=20).

The critical values of the cooperative coefficient C for appearance of the bistable region (Ccrit) and for appearance of the differential amplification region (C'crit) are calculated by the reasoning that at Ccrit the two roots of the equation $\frac{dy}{dx_T}=0$ must coincide and at C'crit the two roots of the equation $\frac{dy}{dx_T}=1$ must coincide. Thus we obtained the following formulae

$$\mathrm{Ccrit} = 2\left(\frac{1}{N} + 1\right) \tag{3}$$

$$\mathrm{C'crit} = 2\left(\frac{1}{N} + 1\right) - \frac{4}{\sqrt{N}} \tag{4}$$

Both Ccrit and C'crit are affected by transmission coefficient ratio N (see Fig.3).

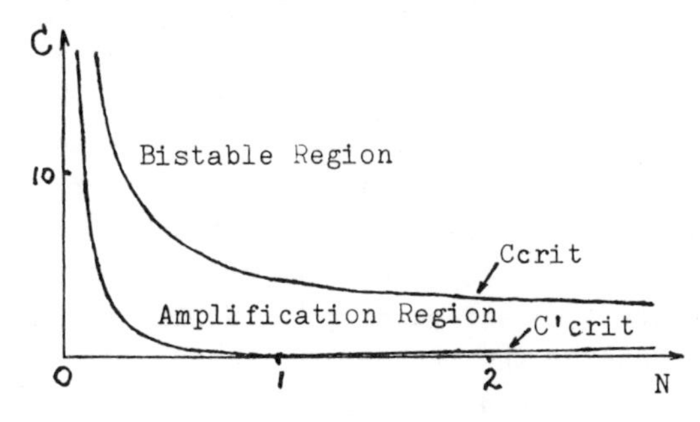

Fig.3. The three different regions in the parameter plane of C and N.

For arbitrary C, the two different roots of the equation $\frac{dy}{dx_T}=0$ can be obtained explicitely and are denoted by x_{T1} and x_{T2}. The corresponding y's obtained from eq.(1) are denoted by y_1 and y_2. Then the height Δx_T^2 and the breadth Δy^2 of the bistable loop are as follows

$$\Delta x_T^2 = x_{T1}^2 - x_{T2}^2 \tag{5}$$

$$\begin{aligned} \Delta y^2 &= y_2^2 - y_1^2 \qquad (6)\\ &= N\left[\frac{1}{2}\left(1+\frac{1}{N}\right)(x_{T1}+x_{T2}) + 2C\left(\frac{x_{T1}}{1+Nx_{T1}^2} + \frac{x_{T2}}{1+Nx_{T2}^2}\right)\right] \\ &\quad \times \left[\frac{1}{2}\left(1+\frac{1}{N}\right)(x_{T2}-x_{T1}) + 2C\left(\frac{x_{T2}}{1+Nx_{T2}^2} - \frac{x_{T1}}{1+Nx_{T1}^2}\right)\right] \end{aligned}$$

where

$$x_{T1}^2 = -\left(\frac{1}{N} - \frac{2C}{1+N}\right) + \frac{1}{1+N}\sqrt{C^2 - 2C\left(1+\frac{1}{N}\right)}$$

$$x_{T2}^2 = -\left(\frac{1}{N} - \frac{2C}{1+N}\right) - \frac{1}{1+N}\sqrt{C^2 - 2C\left(1+\frac{1}{N}\right)}$$

Thus both Δx_T^2 and Δy^2 can be controlled by the transmission coefficient ratio N. (see Fig.4 and Fig.5)

Through linear stability analysis, we got the following equation which can be used to calculate $\mathrm{Re}\lambda$ for determining the steady state

$$(\tilde{\lambda}+i\tilde{\alpha}_n)[(\tilde{\lambda}+1)^2+NX_T^2]$$
$$=-\tilde{K}\left\{\frac{1}{2}\left(1+\frac{1}{N}\right)[(\tilde{\lambda}+1)^2+NX_T^2]+2C\cdot\frac{\tilde{\lambda}+1-NX_T^2}{1+NX_T^2}\right\} \quad (7)$$

where $\tilde{\lambda}=\lambda/\gamma$; $\tilde{\alpha}_n=\alpha_n/\gamma=2\pi nc/\mathcal{L}\gamma$, $n=0,\pm1,\cdots$; $\tilde{K}=K/\gamma=cT1/\mathcal{L}\gamma$.

We argue that although T1 must be small for the validity of the mean field approximation, yet still beside the systems with $\tilde{K}\ll1$, there can be many other systems with $\tilde{K}\gg1$ in the case of a very short cavity or of an atomic material with long relaxation times. We have calculated Reλ for these two classes of systems. For $\tilde{K}\ll1$ cases, the unstable region includes the negative slope section and a part of upper branch of the steady state curve just as L.A.Lugiato et al studied.[1][2] The latter leads to self-pulsing. For $\tilde{K}\gg1$ cases, it turns out that the unstable region coincides with the negative slope section of the steady-state curve. Hence the self-pulsing instability can no longer occure. This may be one of the reasons that the self-pulsing was not often observed in many past experiments.

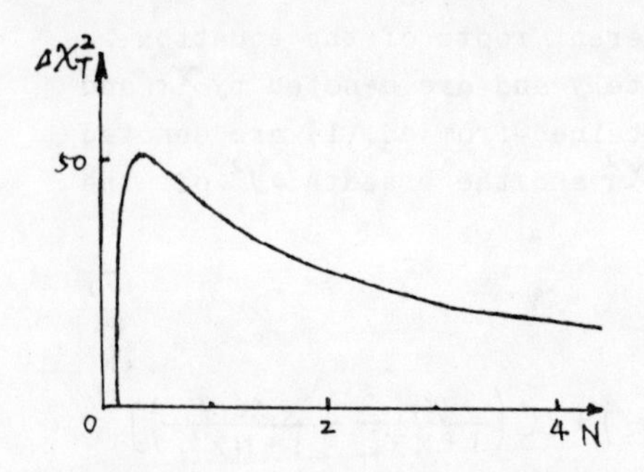

Fig.4. Plot of the height of the bistable loop versus N for C=22.

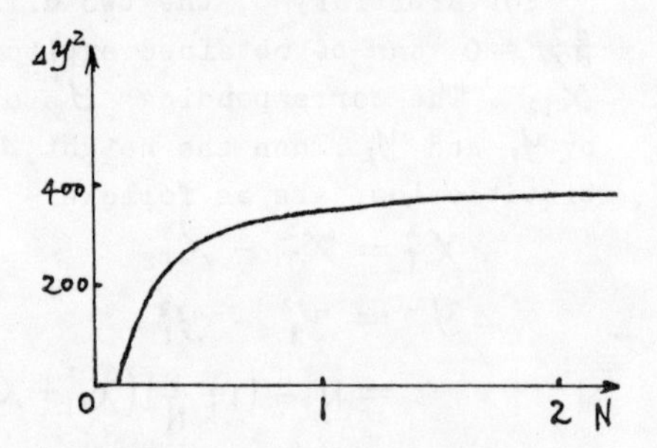

Fig.5.Plot of the breadth of the bistable loop versus N for C=22.

References

1. R.Bonifacio and L.A.Lugiato Lett. Nuovo Cim. 21(1978) 510
2. R.Bonifacio,M.Gronchi and L.A.Lugiato Opt.Commun.30(1979)129
3. L.A.Lugiato, in Progress in Optics, Vol. XXI, ed. by E.Wolf, (North-Holland, Amsterdam 1984) p69
4. L.A.Lugiato, L.M.Narducci Phys. Rev. A32(1985), 1576,1563

A POTENTIAL WELL MODEL OF OPTICAL BISTABILITY

Ai-Qun Ma, Li-Xue Chen and Chun-Fei Li

Department of Applied Physics, Harbin Institute of Technology
Harbin, PRC

Weng Zhao-Heng

Changchun Inst. of Optics and Mechanics, Academia Sinica
Changchun, PRC

We once used catastrophe theory model [1] to study optical bistability and obtained significant results. We also gave a quantum mechanical model [2] to study multi-photon optical bistability and obtained identical results as in reference [3]. With the adiabatic approximation, we obtained the potential well model to study multi-photon optical bistability.

According to the quantum mechanical model of multi-photon transition, we can obtain the differential equation as follows [4]

$$\dot{\alpha} = -(\partial V / \partial \alpha^*) \qquad (1)$$

where V is the potential function of radiation field. V has the following form,

$$V(x, y) = k\, N_{sn} [\, x^2 - 2xy - (1/n)(\eta / \overline{\eta}_n)\ ln(1 + x^{2n})] \qquad (2)$$

$$x_s = \begin{cases} x_1 & (y < \bar{y}) \\ x_3 & (y > \bar{y}) \end{cases} \qquad (3)$$

where x_1 is the point corresponding to cooperation branch and x_3 is the point corresponding to one atom branch, any other point of steady state is inferior stable.

2. The explantion of the switching characteristic.

It is shown clearly in Fig. 1, accompanying the increase of y, the optical bistability system would move from stable region to inferior stable region along the cooperation ditch (It corresponds to cooperation branch). When y is equal to y_2, the system is at unstable state and jumps into the stable region of monoatomic ditch (It corresponds to monoatomic branch) at once. From the curve the system can not maintain at point of inflexion, so it has to enter into the minimum point and the jump process from cooperation branch to monoatomic branch is finished. If y is increased continiously, the system would move along monoatomic ditch, and if y is decreased the system would moving back. When y is equal y_1, the system can not maintain at point of inflexion too, so it has to enter into cooperation ditch. The above description is a deterministic physical explanation to the switching characteristic. The switching characteristic is that the stable state becomes an inferior stable state, then it suddenly changes into the other stable state.

3. The explanation of the hysteresis characteristic.

As shown in Fig. 1, accompanying the change of intensity of incident light the system moves along these two ditches, but these two ditches

Eq.(2) is the analytical experession of potential function of multi-photon bistability system.

In three dimensional space (V, x, y), we make up the three dimensional map of Eq.(2), which is the potential well model of multi-photon optical bistability, as shown in Fig.1. The characteristic of Fig.1 is that it has two "ditches" which are separated by a "mountain range". When multi-photon optical bistability system is at stable state and inferior stable state, it likes a steel ball which is at the bottom of the ditches. The steel ball moves in the ditches according tothe increasing and decrease of the intensity of incident light. When the intensity of incident light is increased or decreased to a certain level, the steel ball can also get into another ditch from one ditch.

The application of the potential well model,

1. The separation of stable region and inferior stable region in the bistability characteristic.

From Fig.1 we find that, only when $y=\bar{y}$, two minimums are the same in all potential function plane curves. The quantum mechanics told us, only when $y=\bar{y}$, the system has bistable states. In other condition, it has only one stable state or it has a stable state and an inferior stable state. So, multi-photon optical bistable states ought to be called multi-photon optical stable-inferior stable two states. According to Fig.1, we can find that the point of steady state of multi-photon optical bistability characteristic which satisfies the condition (3) is absolute stable.

do not touch each other, so multi-photon optical bistability has hysteresis characteristic. We may determine the switching point y_1, y_2 and bistable region (y_1, y_2)using this potential well model.

4. The judgement of the type of optical bistability phase transition.

Landau' s phase transition theory told us the characteristic of phase transition of the first order is the system happens sudden change. According to this potential well model, it is easy to judge that the phase transition happened in optical bistability system is non-equilibrium phase transition of the first order.

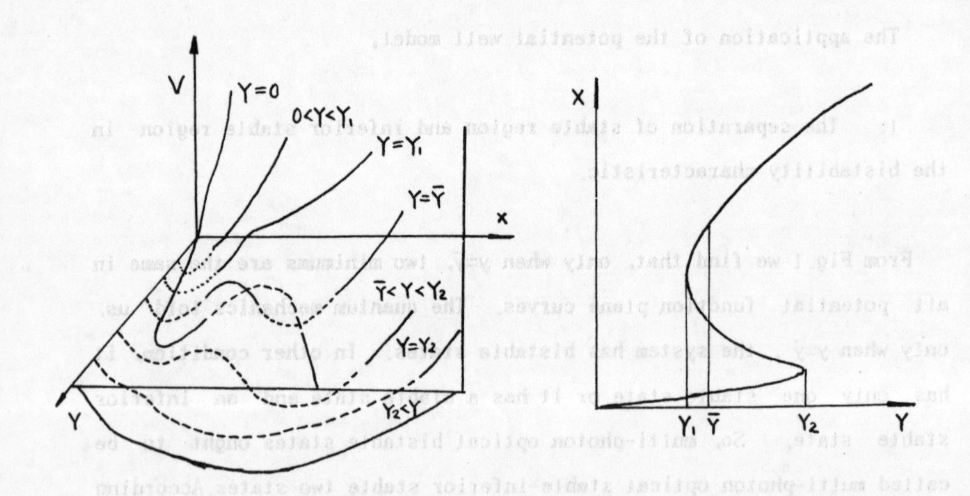

Fig. 1 The potential well model of of optical bistability

Fig. 2 The characteristic of optical bistability

REFERENCES

[1] Chun-Fei Li and Ai-Qun Ma, Acta Optica Sinica, vol.3(1983)811

[2] Weng Zhao-Heng, Ma Ai-Qun and Li Chun-Fei, Acta Optica Sinica, vol.5(1986)441

[3] L.A.Lugiato, et al, Opt. Commun., vol.41(1982)441

[4] H. Haken, 《synergetics》, Springer Verlag Berlin Heidelberg, New York, 1977

OPTICAL BISTABILITY AND SPONTANEOUS SWITCHING BETWEEN STEADY STATE SOLUTIONS IN A BIDIRECTIONAL RING LASER

N.B. Abraham, L.M. Hoffer, and G.L. Lippi, Department of Physics,
Bryn Mawr College, Bryn Mawr, PA 19010 USA
and
H. Zeghlache and P. Mandel, U. Libre Bruxelles, Campus Plaine, CP 231,
Boulevard du Triomphe, B-1050 Bruxelles, Belgium

OVERVIEW

A bidirectional ring laser has four steady state solutions: two unidirectional solutions, the trivial solution, and a bidirectional solution. We have analyzed this problem for a single mode operating in each direction in the laser interacting with the same homogeneously broadened medium. We have adiabatically eliminated the (fast-relaxing) polarization - as is appropriate for YAG, CO_2, and ruby lasers and for some simple models for semiconductor lasers. Under these conditions only the two unidirectional solutions are ever stable above the lasing threshold. Detuning of the laser from resonance can destabilize these states. The amount of detuning required depends on the ratio of the population decay rate to the field decay rate and can be quite small if the ratio is small. In addition, the loss of stability appears for very modest excitation above the laser threshold. A key feature in the dynamics of this system is the formation of a longitudinal spatial grating in the population inversion caused by variations in the gain saturation which result from constructive and destructive interference of the counter-propagating fields. The slow decay rate of the population inversion means that the grating can persist long after the fields which created it have decayed from the cavity. The grating leads to additional coupling between the counter-propagating fields beyond simple cross-saturation of the gain. One field can now scatter from the grating into the other field. However, the grating is time dependent and its dynamics can affect the stability of the lasing operation. With detuning, the grating has a complex (phase-shifted) amplitude, which can make the scattered fields constructively or destructively interfere with the fields that created the grating.

For conditions that destabilize unidirectional cw operation, complex periodic and chaotic pulsations have been observed, but near the instability boundary we have observed irregular (probably chaotic) "square-wave switching" between the two unidirectional steady states - a feature which has also been observed in CO_2 laser experiments.

Closer examination of the seven-dimensional phase space and of the analytical and numerical results reveals that near each steady state solution the trajectories converge in four dimensions and diverge in an oscillatory manner in two dimensions which are "discovered" by the trajectory only after it comes very close to the steady state. Some of the complexity of the laser output and the chaotic pulsations can be traced to the fact that the four-dimensional stable manifold near each unidirectional steady state solution involves two orthogonal two-dimensional manifolds on which the trajectory contracts in spirals. The two contracting spirals have frequencies which differ by the square root of two, and they become coupled when the trajectory leaves the stable manifold in a third outward spiral on the unstable manifold. The coupling of these two incommensurate frequencies in the switching process seems to provide an origin for sensitive dependence on initial conditions, reflected in divergent trajectories and overall chaotic (deterministically irregular) switching.

We also have analyzed the possibility of unequal losses for the two counter propagating fields and present some results for the different stability of the two unidirectional modes in this case.

Our numerical solutions for time-dependent behavior show critical slowing down of the time-dependent solutions at the boundary of the most stable steady-state solution. This suggests that this boundary separates the stable steady state solutions from the stable time-dependent solutions and thus that there are no stable time-dependent solutions coexisting with the stable steady state solutions.

I. EXPERIMENTAL MOTIVATION

Bidirectional ring lasers have been studied for almost the full history of laser physics, particularly because of their applications as gyroscopes. Our particular interest grew out of experimental studies of a CO_2 laser ring at the Istituto Nazionale di Ottica in Florence, Italy in 1983-84[1-3].

Initially we hoped to make the CO_2 laser unidirectional by placing a back-reflecting mirror, M in one of the two output beams as shown schematically in Figure 1.

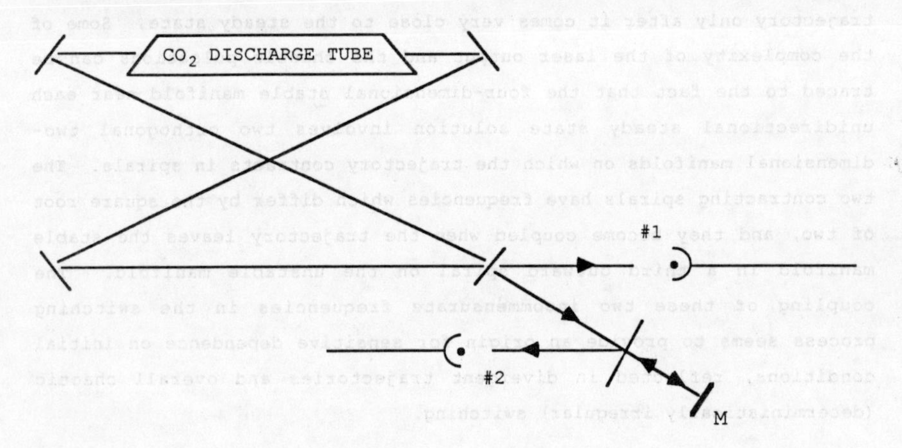

Fig. 1. Schematic of our CO_2 ring laser.

This is commonly reported as being successful in dye lasers[4] in suppressing the second beam by enhancing the first beam. In our preliminary experiments we discovered that the intensity output at detector #1 showed pulsations. The pulsation pattern depended on the position of the external mirror. When we also detected a portion of beam #2, we discovered that it was also pulsing. Furthermore, when the back-reflecting mirror was removed, both beams continued to pulse. Two of the types of pulsations that we observed are shown in Figure 2.

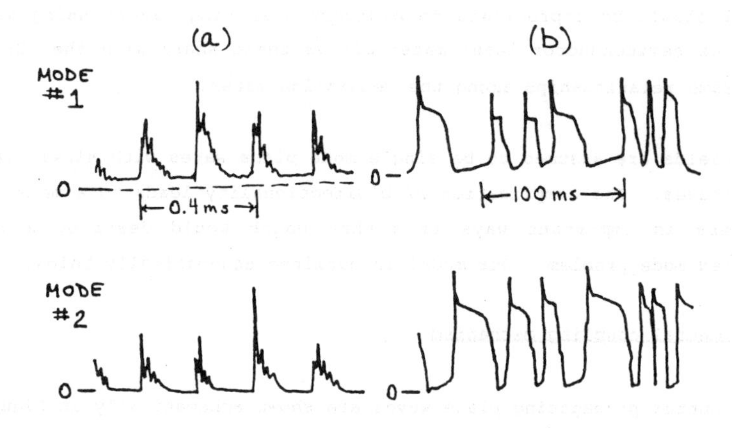

Fig. 2. Sketches of experimental observations of two types of pulsations in the two modes: a) in-phase and b) out-of-phase.

While our attempts to realize a single-mode laser were unsuccessful, we found an interesting two-mode dynamical system on which many experimental measurements have been made[1-3]. These results motivated us to undertake the theoretical analysis of this complex dynamical system.

II. THEORETICAL MODEL

Theoretical studies of a bidirectional ring lasers began with Aronowitz[5] and Agabekyan and coworkers[6]. The extensive literature on this problem has been reviewed recently by Khanin[7] and Khandokhin and Khanin[8] and by us in an earlier report of some of our theoretical results[9].

In a CO_2 laser (of 10-20 Torr) the polarization decay rate is about ten times larger than the decay rate of the field which is itself almost 10^4 times larger than the decay rate of the population inversion. Thus we are confident that, as long as the bandwidth of the pulsation spectrum remains small compared with the decay rate of the polarization, we can adiabatically eliminate the polarization. The model then involves equations only for the fields and the population inversion. Such a model should be appropriate to bidirectional ring lasers using Nd:YAG, ruby or semiconductor laser materials as these share with the CO_2 laser the same relationships among the relaxation rates.

The fields are assumed to be single-mode plane waves with slowly varying amplitudes. The complication of bidirectionality leads to a model which differs in important ways from that which would describe a simple coupled mode problem. Our model is outlined schematically below.

Fundamental coupling mechanism

The counter-propagating plane waves are shown schematically in Figure 3.

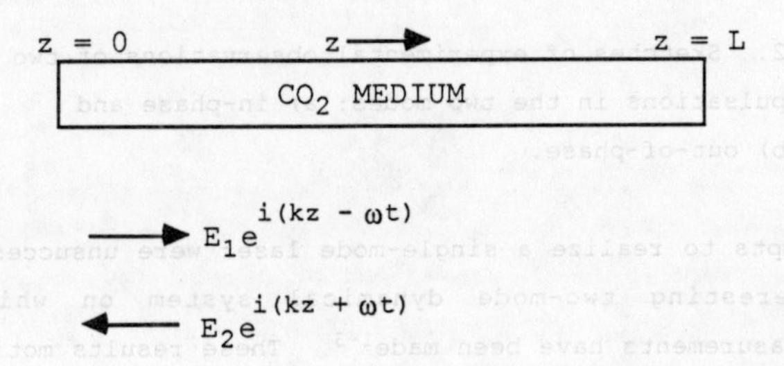

Fig. 3. Counterpropagating waves in the medium.

The intensity varies in space as

$$I(z) = |E_1|^2 + |E_2|^2 + 2 \text{ Re } E_1 E_2^* \cos 2kz. \tag{1}$$

The population inversion per atom may be taken to be spatially uniform in the unsaturated (I = 0) case. In the presence of a stationary field, the inversion is given by

$$D = \frac{D_o}{1 + I/I_s}, \tag{2}$$

where D_o is the unsaturated inversion and I_s is the saturation intensity of the medium.

When $I \neq 0$, D is modified as shown schematically in Figure 4.

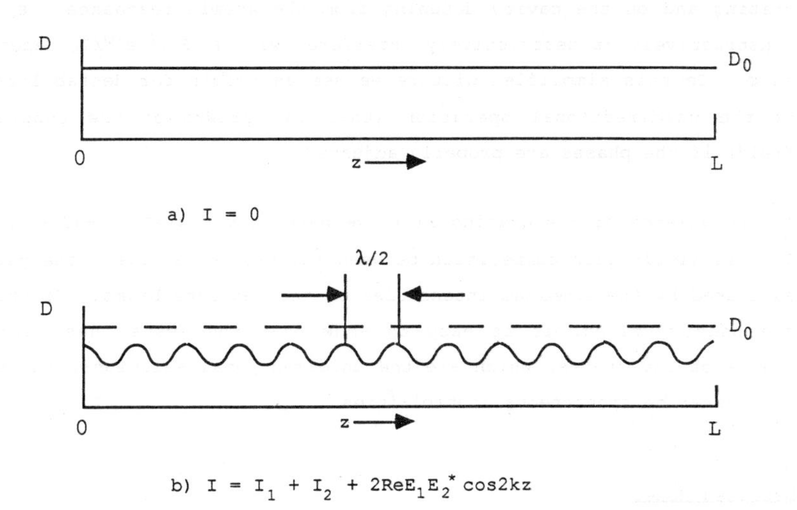

Fig. 4. Longitudinal hole burning in the population inversion.

If the amplitude of the grating is small, we can binomially expand equation (2) keeping only the harmonic term. Then

$$D(z) = D(0) + D(1) \cos 2kz, \tag{3a}$$

where

$$D(0) = \frac{D_o}{1 + (I_1+I_2)/I_s}, \text{ and} \tag{3b}$$

$$D(1) = - \frac{D_o}{1 + (I_1+I_2)/I_s} \; \frac{2 \, \mathrm{Re} E_1 E_2^* / I_s}{1 + (I_1+I_2)/I_s} \; . \tag{3c}$$

The infinite harmonic sequence from the binomial expansion truncates rapidly ($|D(1)| \ll D(0)$) if $|E_1 E_2^*| \ll I_s$.

The spatial grating scatters E_1 into E_2 and vice-versa. In the presence of only E_1 with $E_2 = 0$ we see that a small fluctuation, ε_2, in E_2 creates a grating of amplitude $\sim|\varepsilon_2 E_1^*|/I_s$ which will then scatter some of E_1 into E_2 in the amount $E_1\varepsilon_2 E_1^* e^{i\varphi}/I_s$, where φ depends on the phase of the grating and on the cavity detuning from the atomic resonance. ε_2 then constructively or destructively interferes with $\varepsilon_2|E_1|^2 e^{i\varphi}/I_s$ depending on φ. In this simplified picture we see an origin for destabilization of the unidirectional operation (that is, growth of the suppressed field) if the phases are properly adjusted.

In the absence of the grating D(1), we have only $D(0)=D_o/[1+(I_1+I_2)/I_s]$. This is simply gain competition between the two beams, i.e., the gain is saturated by the combined intensities of the separate beams. If this is the only coupling, it is easy to show that the system has only two stable steady states, which are the unidirectional solutions, and there will never be spontaneous instabilities.

Detailed Model

The formal model of the system is given as follows:

$$(\partial_\tau+1)E_1 = \tilde{A}E_1 D(0,\tau)\ (1+i\Delta) + \tilde{A}E_2 D^*(1,\tau)\ (1+i\Delta) \tag{4a}$$

$$(\partial_\tau+K)E_2 = \tilde{A}E_2 D(0,\tau)\ (1+i\Delta) + \tilde{A}E_1 D(1,\tau)\ (1+i\Delta) \tag{4b}$$

$$(\partial_\tau+d_{||})D(0,\tau) = d_{||} - \tilde{d}_{||}D(0,\tau)\ [|E_1|^2 + |E_2|^2] - \tilde{d}_{||}[D(1,\tau)E_1E_2^* + D^*(1,\tau)E_1^*E_2] \tag{4c}$$

$$(\partial_\tau+d_{||})D(1,\tau) = -\tilde{d}_{||}D(1,\tau)[|E_1|^2 + |E_2|^2] - \tilde{d}_{||}D(0,\tau)E_1^*E_2 \tag{4d}$$

where the parameters are as follows:

$d_{||} \equiv \gamma_{||}/\gamma_{\perp}$,

$K \equiv \dfrac{\kappa_2}{\kappa_1}$,

$\Delta \equiv (\omega_c - \omega_A)/\gamma_{\perp}$,

$\tilde{d}_{||} \equiv d_{||}/(1+\Delta^2)$,

$\tilde{A} \equiv A/(1+\Delta^2)$,

$\gamma_{||}$ is the population inversion decay rate,

$\gamma_{\perp}$ is the polarization decay rate

κ_1 is the field decay rate for mode #1,

κ_2 is the field decay rate for mode #2,

A is the excitation parameter depending on the number of excited atoms scaled to be 1 at the lasing threshold for mode #1,

D is the population inversion of a single atom,

E has been scaled to the square-root of the saturation intensity,

D(0) is a real variable, E_1, E_2 and D(1) are complex variables,

τ is time normalized to the decay rate of the field of mode #1 $\tau = \kappa_1 t$,

ω_c is the resonant frequency of the laser cavity,

ω_A is the resonant frequency of the medium.

III. STEADY-STATE SOLUTIONS

The equations (4) have four steady state solutions. These are given by:

i) $I_1 = 0;\ I_2 = 0$

ii) $I_1 = A-(1+\Delta^2);\ I_2 = 0$

iii) $I_1 = 0;\ I_2 = A-K(1+\Delta^2)$

iv) $I_1 \neq 0;\ I_2 \neq 0$

and they can be pictured in the partial phase space (I_1, I_2) as shown in Figure 5.

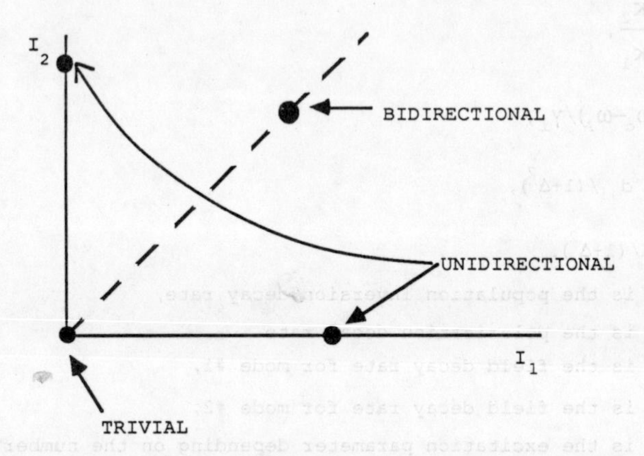

Fig. 5. Steady state solutions in phase space (I_1, I_2).

Above the lasing threshold the trivial solution is unstable so we will generally not be concerned with it. Similarly, the bidirectional solution is always unstable[10]. In resonance, for most excitation levels, the other two solutions are stable, but they can be destabilized by detuning of the cavity from resonance.

> [Aside: It may be that further studies should consider the relative instabilities of all four steady state solutions in the domain where all are unstable. There may be interesting relevance of the trivial solution or the bidirectional solution to the dynamics if their degrees of instability are comparable to the degree of instability of the unidirectional solutions.]

When $K > 1$, mode #1 has less loss than mode #2. As a result the intensity of the $I_1 \neq 0$ unidirectional solution is larger than that of the $I_2 \neq 0$ unidirectional solution. Therefore, we refer to the former as "the strong mode steady state solution" and to the latter as "the weak mode steady state solution".

IV. STABILITY ANALYSIS

The linearized stability analysis of the seven-equation model in the vicinity of the unidirectional steady states solutions gives the following results. The stability boundary of the weak mode solution is always contained within the stability of the strong mode solution as shown for example in Figure 6.

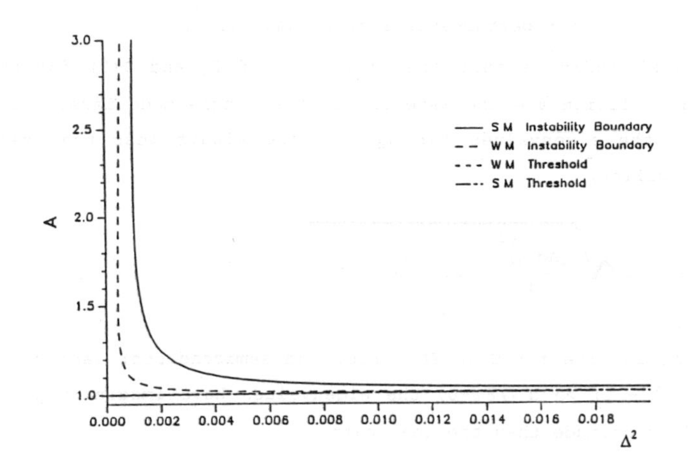

Fig. 6. Stability boundaries in the parameter space (A, Δ^2) for the weak and strong modes for K=1.00007, $d_{||}=1.8 \times 10^{-4}$. Weak mode and strong mode thresholds cannot be resolved on this graph.

There are three regions:

I) BISTABLE - both unidirectional steady state solutions are stable.

II) MONOSTABLE - only the strong mode unidirectional steady state solution is stable.

III) UNSTABLE - there are no stable steady state solutions.

The stability analysis of the strong mode steady state solution gives a seventh order characteristic equation which can be factorized as follows:

$$\lambda = 0 \qquad (5a)$$

for perturbations of the phase of E_1,

$$\lambda^2 + \lambda d_{||}\tilde{A} + 2d_{||}(\tilde{A}-1) = 0 \qquad (5b)$$

for perturbations of $|E_1|$ and D(0),

$$\lambda^2 + \lambda(d_{||}\tilde{A}+(K-1)) + d_{||}(K\tilde{A}-1) + i\Delta d_{||}(\tilde{A}-1) = 0 \qquad (5c)$$

for perturbations of E_2 and D(1), and

$$\lambda^2 + \lambda(d_{||}\tilde{A}+(K-1)) + d_{||}(K\tilde{A}-1) - i\Delta d_{||}(\tilde{A}-1) = 0 \qquad (5d)$$

for perturbations of E_2^* and $D(1)^*$.

Equations (5a,b) indicate that the stability of E_1 and D(0) for the unidirectional solution are the same as for the single mode laser. The real parts of the eigenvalues for Eq. (5b)are always less than zero indicating stability,

$$\lambda = -\frac{\tilde{A}d_{||}}{2} \pm \sqrt{\frac{(\tilde{A}d_{||})^2}{4} - 2d_{||}(\tilde{A}-1)} \quad . \qquad (6)$$

If $d_{||} << 1$, as is true for CO_2, YAG, ruby and semiconductor lasers, we see that for $(\tilde{A} - 1) >> d_{||}/8 \sim 0$, the eigenvalues have imaginary parts much larger in magnitude than the real part

$$\lambda \cong -\frac{\tilde{A}d_{||}}{2} \pm i\sqrt{2d_{||}(\tilde{A}-1)} \quad , \qquad (7)$$

indicating that the system will relax along these coordinates with weakly damped oscillations--a spiral in the $|E_1|$, D(0) plane. In physical units the eigenvalue is given by

$$\hat{\lambda} \cong -\frac{\tilde{A}\gamma_{||}}{2} \pm i\sqrt{2\kappa_1\gamma_{||}(\tilde{A}-1)} \; . \qquad (8)$$

For Eq. (5c) the eigenvalues are given approximately by

$$\lambda_{1,2} \cong -\frac{\left[d_{||}\tilde{A} + K - 1\right]}{2} \pm i\left[1 - i\frac{\Delta}{2}\right]\sqrt{d_{||}(\tilde{A}-1)} \; , \qquad (9)$$

if $d_{||} << 1$, $K - 1 << 1$ and $\Delta << 1$. For Eq. (5d) the eigenvalues are the complex conjugates of those given in Eq. (9). There are two stable directions in the combined E_2, E_2^*, $D(1)$, $D(1)^*$ variable space corresponding to λ_2. Here again we have an inward spiral of weakly damped oscillations, but the frequency differs from that of Eqs. (7) and (8) by $\sqrt{2}$.

In the other two directions, stability is governed by $\mathrm{Re}\lambda_1$.

$$\mathrm{Re}\lambda_1 = -\frac{[d_{||}\tilde{A} + K-1]}{2} + \frac{\Delta}{2}\sqrt{d_{||}(\tilde{A}-1)} \ . \tag{10}$$

$\mathrm{Re}\lambda_1 = 0$ at approximately

$$\Delta_c \approx \frac{[d_{||}\tilde{A}_c+K-1]}{\sqrt{d_{||}(\tilde{A}_c-1)}} \ . \tag{11}$$

More precisely we can find the instability point $\mathrm{Re}\lambda_1 = 0$ by substituting $\lambda_1 = i\Omega$ (Ω real) into Eq. (5c), yielding

$$-\Omega^2 + i\Omega\,(d_{||}\tilde{A} + K-1)) + d_{||}(K\tilde{A}-1) + i\Delta d_{||}(\tilde{A}-1) = 0. \tag{12}$$

The real and imaginary parts of Eq. (12) satisfy

$$\Omega^2 = d_{||}(K\tilde{A}_c-1) \tag{13a}$$

$$\text{and } \Omega = -\frac{\Delta_c d_{||}(\tilde{A}_c-1)}{d_{||}\tilde{A}_c + K-1} \tag{13b}$$

$$\text{so } \Delta_c^2 = \frac{(K\tilde{A}_c-1)\ [d_{||}\tilde{A}_c + K-1]}{d_{||}(\tilde{A}_c-1)^2} \tag{14}$$

$$\text{and } \Omega = \sqrt{d_{||}(K\tilde{A}_c-1)} \ . \tag{15}$$

Eq. (14) governs the boundary of stability of the strong mode solution as plotted in Figs. 6 and 7. Asymptotic analysis gives the characteristic features of this curve for $d_{||} << 1$ as labelled in Fig. 7.

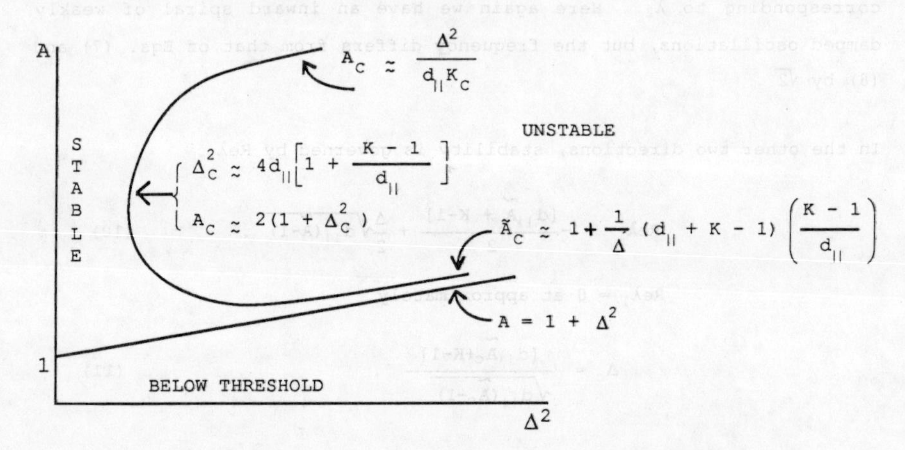

Fig. 7. Asymptotic limits of stability boundary for the strong mode.

The region near threshold and near $\Delta = 0$ can be expanded for the case of $K \neq 1$ as shown in Fig. 8.

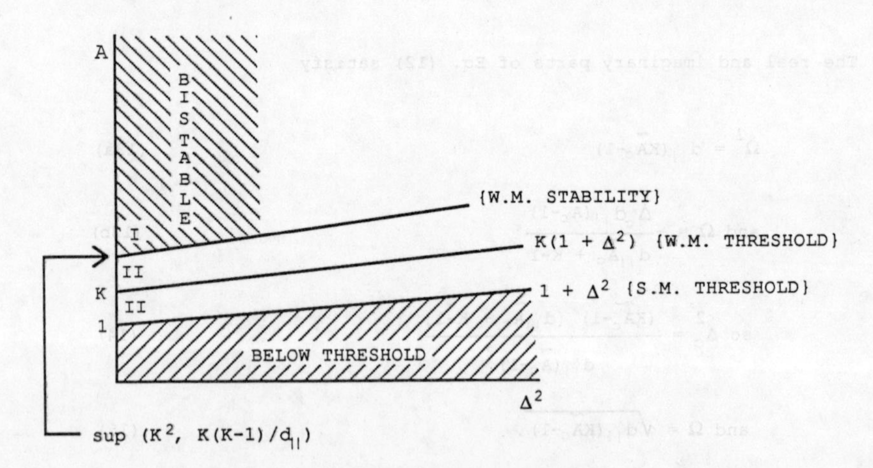

Fig. 8. Expanded view of Fig. 6 near Δ=0 and A=1. I, bistable; II, only the strong mode unidirectional solution is stable.

From these various results we see that the laser is unstable for both symmetric losses (K = 1) and asymmetric losses (K > 1). Asymmetric losses lead to a wider monostable range for $\Delta = 0$ as A is increased. Even when A is above the threshold for the weak mode, the weak mode does not gain stability until A is raised above $\sup(K^2, K(K-1)/d_{||})$. For example, if K = 1.1 and $d_{||} = 10^{-4}$, the monostable region exists for $1 < A < 10^3$. For K = 1.0001 and $d_{||} = 10^{-4}$, the monostable region exists for

$$1 < A < 1.0002.$$

To achieve a significant region of stable single mode operation at $\Delta=0$, it is sufficient to raise the difference in the losses ($K - 1 \gg d_{||}$). Since for our lasers $d_{||} \cong 10^{-4}$, only a small mismatch in losses can eliminate the bistable region from experimental accessibility.

Many authors have noted that linear optics do not permit a difference in losses for the two modes. However an equivalent loss differential can be provided by a small Faraday effect or by nonlinear transverse effects related to gain and dispersion focussing and saturation which are simulated in plane wave theories by differences in the losses[11].

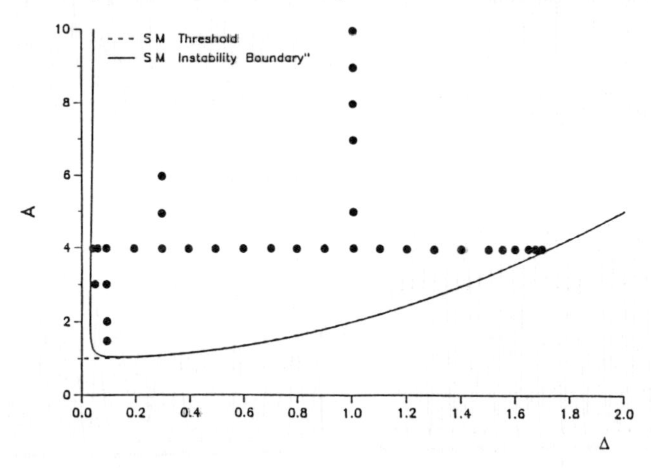

Fig. 9. Region explored for CO_2 laser numerical solutions with K=1.00007 and $d_{||}=1.8 \times 10^{-4}$.

V. TIME-DEPENDENT SOLUTIONS

Numerical solutions of Eqs. (4) were calculated for parameters selected to match the CO_2 laser experiments. The portion of the phase space explored thus far is shown in Figure 9. Samples of periodic, complex chaotic, and nearly square-wave chaotic solutions are shown in Figure 10.

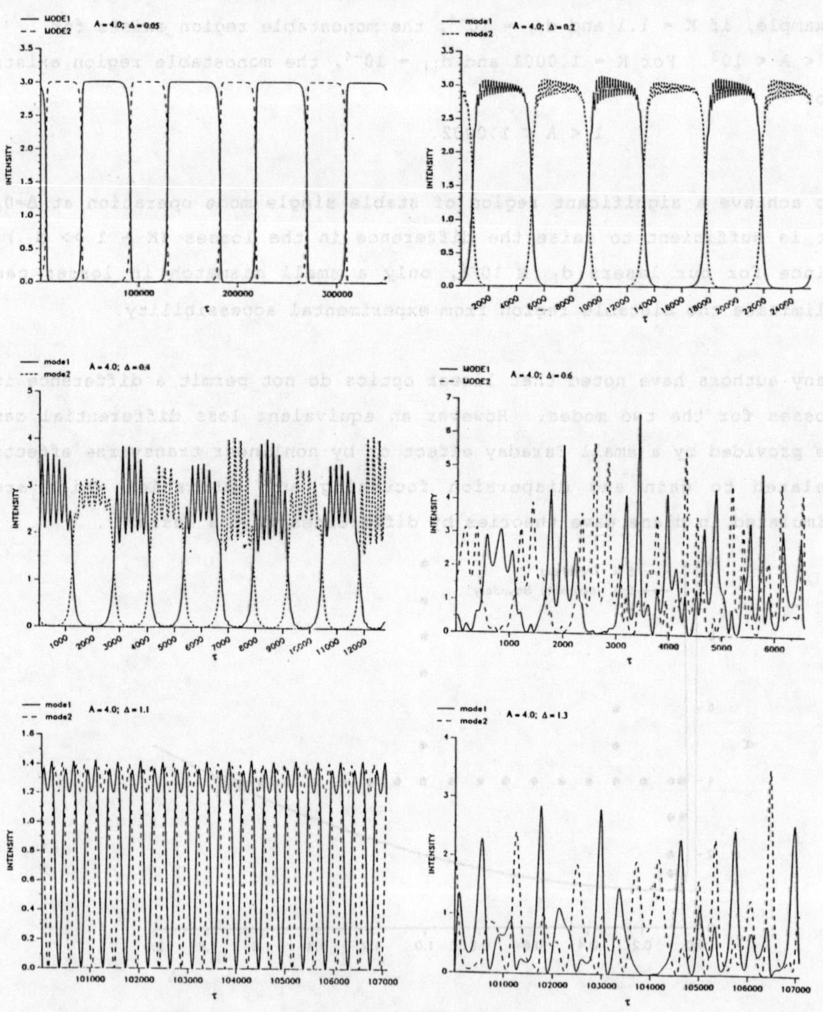

Fig. 10. Time dependent solutions.

The "square wave chaotic" switching occurs close to the boundary. The details of the switching process are shown in Figure 11.

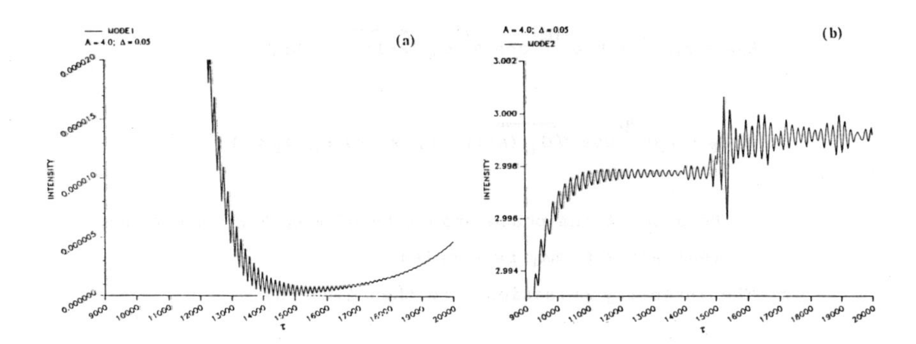

Fig. 11. Expanded view of a "square wave switch".

We see that the solution relaxes toward $I_1 = A-1-\Delta^2 = 2.9975$ and $I_2 = 0$ (or, alternately, to the other unidirectional state) with spirals (leading to intensity pulsations) that differ in frequency by $\sqrt{2}$ with the higher frequency corresponding to the modulation of I_1.

When the intensity of I_2 nearly reaches zero, a second, incommensurate modulation of I_1 begins. The switch reversal starts with quasiperiodic modulation in I_1 and exponentially growing intensity in I_2 with weakening in its modulation.

VI. DISCUSSION

Relation of Pulsation Frequencies to Eigenvalues.

Recall that the spiraling frequencies (when $K \cong 1$) from the linear stability analysis were $\sqrt{2d_{||}(\tilde{A}-1)}$ for the strong field and $\sqrt{d_{||}(\tilde{A}-1)}$ for both the stable and unstable spirals of the weak field. The prediction is that the spiraling frequency of the strong mode should be $\sqrt{2}$ larger than those of the weak mode. The numerical result shown in Fig. 11 of a higher pulsing frequency for the weak mode at first glance appears to contradict the theoretical prediction. We can resolve the discrepancy as follows:

a) The spiraling predictions are for the fields.

b) Hence for the collapsing spirals

$$E_{ON} = E_{ON}^{ss} + \varepsilon_1 e^{-\alpha t} \cos \sqrt{2d_{||}(\tilde{A}-1)}\; t, \text{ and}$$

$$E_{OFF} = \varepsilon_2 e^{-\beta t} \cos \sqrt{d_{||}(\tilde{A}-1)}\; t, \text{ where } \varepsilon_1, \varepsilon_2 \ll 1 .$$

Because of the phase stability of E_{ON}, E_{ON}^{SS} and ε_1 are real while ε_2 may be complex.

c) The intensity pulsations are then given by

$$I_{ON} \approx I_{ON}^{ss} + 2\varepsilon_1 e^{-\alpha t} \cos\sqrt{2d_{||}(\tilde{A}-1)}\; t$$

and

$$I_{OFF} \approx |\varepsilon_2|^2 e^{-2\beta t} \cos^2 \sqrt{d_{||}(\tilde{A}-1)}\; t,$$

so

$$I_{OFF} \approx \frac{1}{2} |\varepsilon_2|^2 e^{-2\beta t} + \frac{1}{2}|\varepsilon_2|^2 e^{-2\beta t} \cos\sqrt{4d_{||}(\tilde{A}-1)}\; t.$$

d) In this way we see that I_{OFF}, the intensity of the near-zero mode, has a pulsation frequency that is $\sqrt{2}$ times that of I_{ON}.

Furthermore we note that the modulation amplitude of I_{OFF} seen in Fig. 11a continues to decrease even as the average value of I_{OFF} begins its exponential rise away from zero. We speculate that the collapsing spiral in E_{OFF}, E_{OFF}^*, D(1), $D^*(1)$ is aligned along $\mathrm{Re}E_{OFF}$ - $\mathrm{Im}E_{OFF}$ with a spiral in the ($|E_{OFF}|$,$|D(1)|$) plane. This would give intensity pulsations. While in contrast it appears that the expanding spiral lies in the ($\mathrm{Re}E_{OFF}$,$\mathrm{Im}E_{OFF}$) plane. In this case the expanding spiral gives only exponential growth of the intensity without intensity modulation. We will be testing these conjectures by determining the eigenvectors corresponding to the eigenvalves we have calculated.

Stability Boundary of Time-Dependent Solutions.

As we approach the stability boundary of the strong mode in the phase space of Figure 9 the square-wave switching rate becomes infinite. As a test of critical slowing down at the boundary, we plot the mean switching rate versus the distance from the boundary. The result is plotted in Figure 12.

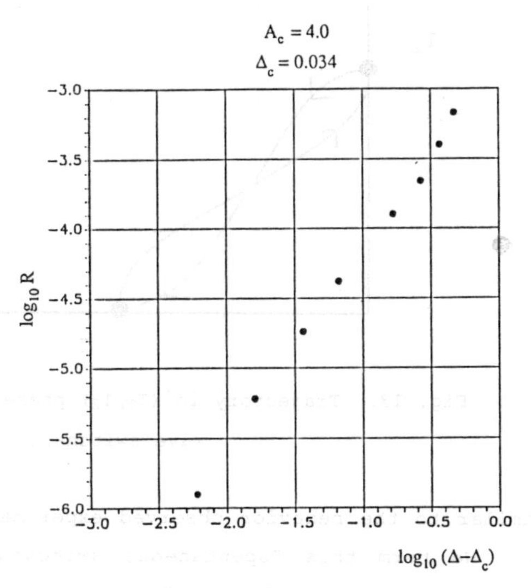

Fig. 12. Log-log plot of switching rate versus distance from the boundary.

The power law dependence with critical exponent 1.6 indicates that the stability boundary of the strong mode steady state is also a boundary of stability for the time-dependent solutions. The specific value of the exponent is larger than the dependence of $\mathrm{Re}\lambda_1$ on Δ from Eq. (9) where the exponent is one. This may be due to the additional dwell time in the vicinity of the steady-state solutions.

It is not common that large amplitude time-dependent solutions share a stability boundary with the stability boundary of the steady state solutions subject to infinitesimal perturbations. It appears that it happens in this case because the large amplitude solutions closely approach one unidirectional steady state solution before switching to the other.

We can see the switching between steady states in Fig. 13 where the trajectory of the square-wave solution is plotted in the (I_1, I_2) space.

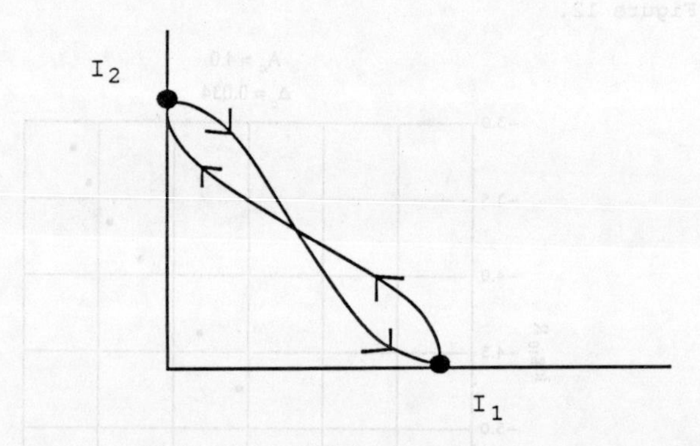

Fig. 13. Trajectory in (I_1, I_2) phase space of square-wave switches.

This is similar to the behavior observed experimentally and shown in Figure 2b. We term this "spontaneous switching between unstable unidirectional steady-state solutions."

The behavior observed experimentally shown in Figure 2a has not been observed in our numerical simulations. Its trajectory would correspond to the motion in the (I_1, I_2) space shown in Figure 14.

Such pulsing may yet be observed in our model, perhaps if the symmetric states are less unstable than the asymmetric states or if there is a suitable difference in losses for the two modes. Alternatively, it may be that this experimental observation corresponds to incomplete excitation of the CO_2 gas, leaving some in the absorbing state. Lasers with saturable absorbers (including a CO_2 laser with a hot CO_2 absorption cell) have shown spontaneous oscillations and auto Q-switching of this type in single mode operation[12,13]. It is possible

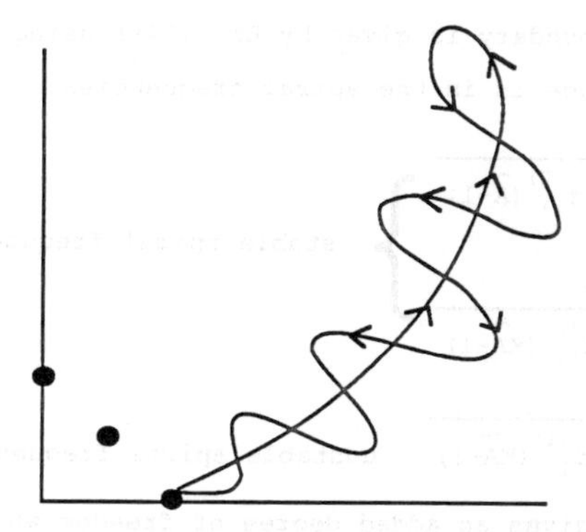

Fig. 14. Trajectory in (I_1, I_2) phase space of in-phase spontaneous Q-switching.

that auto-Q-switching would affect both modes symmetrically. The out-of-phase ringing may arise from complex eigenvalves in the linear stability analysis of the symmetric solutions.

It may also be that symmetric time-dependent solutions will be found if we explore more closely the K = 1 region of parameter space. We have not studied this exactly symmetric condition in our initial numerical studies in order to avoid spurious (or singular) effects.

Effects of diffusion

Since in many lasers there is diffusion of the atoms or molecules which tends to reduce the spatial inhomogeneities in the population inversion, it is of interest to consider different relaxation rates for D(0) and D(1), $d_{||}(0)$ and $d_{||}(1)$, respectively. The stability analysis is straightforward. It is easy to show that Eqs. (5) become

$$\lambda = 0, \tag{16a}$$

$$\lambda^2 + \lambda d_{||}^{(0)}\tilde{A} + 2d_{||}^{(0)}(\tilde{A}-1) = 0, \tag{16b}$$

$$\lambda^2 + \lambda(d_{||}^{(1)}\tilde{A} + K - 1) + d_{||}^{(1)}(K\tilde{A}-1) + i\Delta d_{||}^{(1)}(\tilde{A}-1) = 0, \tag{16c}$$

$$\lambda^2 + \lambda(d_{||}^{(1)}\tilde{A} + K - 1) + d_{||}^{(1)}(K\tilde{A}-1) - i\Delta d_{||}^{(1)}(\tilde{A}-1) = 0. \tag{16d}$$

The stability boundary is given by Eq. (14a) using $d_{||}^{(1)}$. The significant change is in the spiral frequencies:

$$\left.\begin{array}{l}\sqrt{2d_{||}^{(0)}(\tilde{A}-1)} \qquad (17a)\\ \\ \sqrt{d_{||}^{(1)}(K\tilde{A}-1)} \qquad (17b)\end{array}\right\}\ \text{stable spiral frequencies}$$

$$\sqrt{d_{||}^{(1)}(K\tilde{A}-1)} \quad \text{unstable spiral frequency.} \tag{17c}$$

Diffusion thus gives an added degree of freedom which permits adjustment of the spiral frequencies and their ratio. If the particular values or their incommensurability is critical to achieving chaotic or periodic dynamics we may find qualitatively new results in numerical solutions if we vary $d_{||}^{(1)}$. Of particular interest may be the "resonant" case of

$$d_{||}^{(1)}\ (K\tilde{A}-1) = 2d_{||}^{(0)}\ (\tilde{A}-1)\quad . \tag{18}$$

In studying diffusion we must consider the differences in the diffusive processes in gases and semiconductor materials, and the corresponding effects on the relaxation rates. For example, in gas lasers, thermal motion leads to Doppler-shifted resonance frequencies (changes in detuning) as well as moving gratings and both effects may be relevant. These and other questions will be addressed in our continuing studies.

Inhomogeneous broadening of the resonance frequencies for a stationary medium (no diffusion) has already been considered in ref. 9.

VII. CONCLUSION

The bidirectional ring laser provides a rich variety of time-dependent solutions, many of which can be understood physically as arising from switching between the vicinities of the unstable steady state solutions. Periodic and more complex chaotic solutions are observed in other regions of the phase space.

Further topics under investigation include the stability of the symmetric steady states, the search for nearly symmetric time-dependent solutions in the model, investigation of the effect of different relaxation rates for D(1) and D(0) on the time dependent solutions, and a careful study of the time dependent solutions for K = 1; and K-1 >> $d_{||}$.

VIII. ACKNOWLEDGEMENTS

We wish to thank F.T. Arecchi and J.R. Tredicce for very useful discussions about the CO_2 laser operation. T. Mello was most helpful in setting up and debugging the initial numerical integration programs. This work was supported in part by US-NSF Grant Nos. INT83-00149 and ECS82-100263 and by U.S. Army Research Office Contract No. 22919-PH. NBA acknowledges the support of an Alfred P. Sloan Research Fellowship and an OSA travel grant during portions of this work. GLL acknowledges a Borsa di Studio (203.02.17) from the Consiglio Nazionale delle Ricerche of Italy. PM is Senior Research Associate with the FNRS (Belgium). The research in Brussels was supported in part by the Association Euratom-État Belge, a grant from the European Communities, and a grant from the Algerian government.

IX. REFERENCES

1. J.R. Tredicce, G.L. Lippi, F.T. Arecchi and N.B. Abraham, Phil. Trans. Royal Society, London, A, *313*, 411 (1984).
2. G.L. Lippi, J.R. Tredicce, F.T. Arecchi and N.B. Abraham, Opt. Commun. *53*, 129 (1985).
3. G.L. Lippi, N. Ridi, J.R. Tredicce, F.T. Arecchi, and N.B. Abraham, in *Atti del IV Congresso Nazionale di Elettronica Quantistica e Plasmi*, Certosa, Capri 1984 (ENEA, Serie Simposi, 1984), p. 251.

4. V.V. Antsiferov, G.V. Krivoschchekov, V.S. Pivtsov and K.G. Folin, Zh. Tekh. Fiz. *39,* 931 (1969) [Sov. Phys. Tech. Phys. *14,* 696 (1969)]; A.R. Clobes and M.J. Brienza, Appl. Phys. Lett. *21,* 265 (1972); L.S. Kornienko, N.V. Kravtsov and A.N. Shelaev, Opt. Spektrosk. *35,* 775 (1973) [Opt. Spectrosc. *35* 449 (1973)]; G. Marowsky and K. Daufmann, IEEE J. Quantum Electron. *QE-9,* 305 (1977).

5. F. Aronowitz, Phys. Rev. *139,* 635 (1965); and in *Laser Applications,* Vol. 1, M. Ross, ed. (Academic Press, NY, 1971), p. 133.

6. A.S. Agabekyan, A.Z. Grasyuk, I.G. Zubarev, V.I. Svergon and A.N. Oraevskii, Radiotekh. i Electron. *9,* 2156 (1964).

7. Ya.I. Khanin, in *Optical Instabilities,* R.W. Boyd, M.G. Raymer and L.M. Narducci, eds., (Cambridge U. Press, Cambridge, 1986) p. 212.

8. P.A. Khandokhin and Ya.I. Khanin, J. Opt. Soc. Am. B *2,* 226 (1985).

9. H. Zeghlache, P. Mandel, N.B. Abraham, L.M. Hoffer, G.L. Lippi and T. Mello, Phys. Rev. A (to be published).

10. See arguments in ref. 9 and the analysis for $\Delta = 0$ by P. Mandel and G.P. Agarwal, Opt. Commun. *42,* 269 (1982).

11. See detailed discussion of this point and associated refs. 19, 76, 91-99 in Ref. 9.

12. See, for example, E. Arimondo, F. Casagrande, L.A. Lugiato and P. Glorieux, Appl. Phys. B *30,* 57 (1983); E. Arimondo, P. Bootz, P. Glorieux and E. Menchi, J. Opt. Soc. Am. B *2,* 193 (1985); M. Tachikawa, K. Tanii and T. Shimizu, J. Opt. Soc. Am. B *4,* 387 (1987); and references therein.

13. J. Dupré, F. Meyer and C. Meyer, Rev. Phys. Appl. *10,* 285 (1975); R. Ruschin and S.H. Bauer, Chem. Phys. Lett. *66,* 100 (1979); R. Ruschin and S.H. Bauer, Appl. Phys. *24,* 45 (1981).

The Research Work on Optical Computing, Optical Bistability and Chaos in Institute of Physics, Academia Sinica

G. Z. Yang and H. J. Zhang

Institute of Physics, Academia Sinica

1. The Optical General Transformation

Theory

In 1975, Y.P.Huo, G.Z.Yang et al. presented a generalized theory that an arbitrary linear transformation (LT) can be performed by using a pure-phase masks system[1]. The theory proved that, for any given linear unitary transform there is a coherent system composed of holographic masks to realize it with sufficiently high precision if the number of masks is large enough. Also,in the papers a iterative method to design the masks by computer was given.

In recent years, this theory has made a marked progress in reducing the number of masks used in a transform system. G. Z. Yang proposed that a LT can be performed by using single holographic mask[2]. According to this theory for achieving a LT with sufficiently high precision, the amplitude-phase distribution of mask can be uniquely determined by solving a set of equation, as long as the number of sampling points in the mask is equal to product of the number of sampling points in input and output plane .i.e.

$$N=N1\times N2$$

Experiments

1> 8-sequence Walsh Transform (WT)

The primary experiment is using multiple pure-phase masks in a coherent system to realize 8-sequence WT in 1-demensional space. Y.S.Chen et al. designed and established a coherent system composed of 4-masks. It is shown in Fig.1, where

M-mask,

Go-propagation matrix in Fresnal approximation,

F=Wf,

W=M4.Go.M3.Go.M2.Go.M1.Go.

The experimental results are in agreement with theoretical pridiction[3].

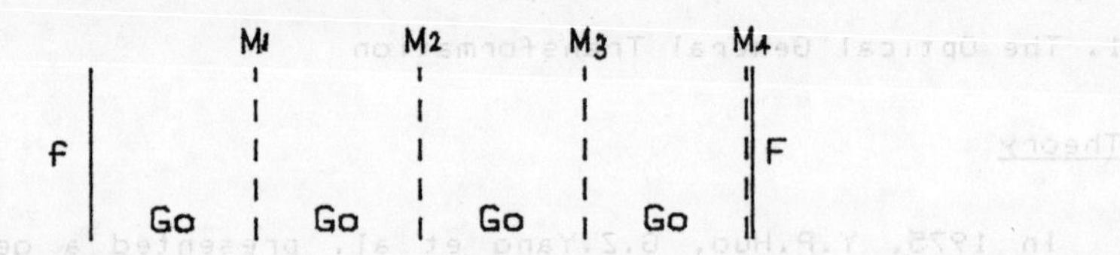

Fig.1. Schematic of 4-mask system.

2) 16--sequence Mellin Transform (MT)

A single mask system was set up by S.H.Zheng et al. to perform the 16-sequence MT in 1-dimensional space. In Fig.2, H is a computer generated hologram[4]

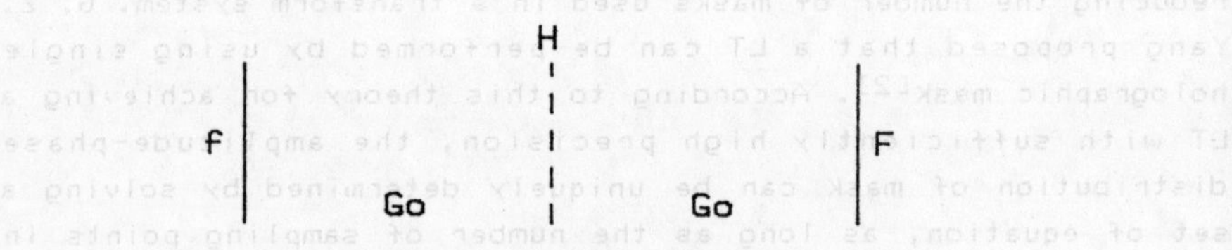

Fig.2 Schematic of single mask system

3) 32-sequence Walsh-Hadamard Transform (SWHT)

Doing experimental research on 32-SWHT with a single holographic mask system is more difficult than in low-sequence (below 16). By a coherent system (shown in Fig.3) consisted of a single mask and two Fourier lenses was presented by Y.S.Chen et al. for solving these problems, such as applicability of paraxial approximation. With this system the 32-SWHT is performed in high precision. The experimental results are consistent with the theoretical expectations.

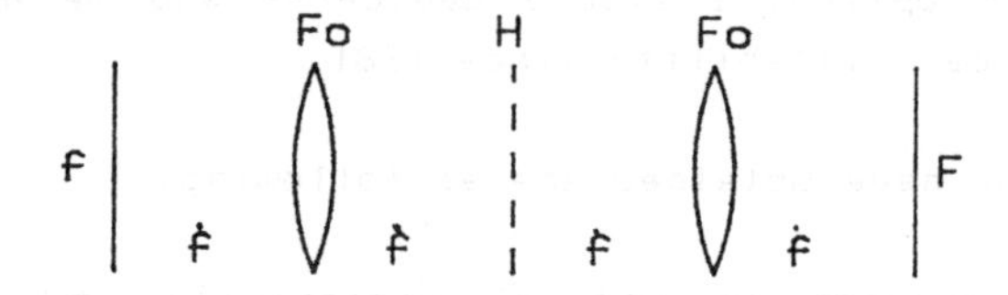

Fig. 3 schematic of system with a single mask and two Fourier lenses

2. Optical Bistability in MQWS Waveguide

Optical bistable devices may someday be the key elements of all-optical logic and computing systems. Semiconductor devices appear particularly promising because of very large optical nonlinearities permitting the construction of very small devices.

Optical bistability using GaAs/AlGaAs multiple quantum well stuctures (MQWS) as the nonlinear element in the Fabry-Perot etalon, has been demonstrated by Gibbs et al. in 1982. In such a device GaAs substrate must be etched out and dielectric coatings are then deposited on both surfaces to increase the reflectivity to nearly 90% between 820 and 890 nm. Such a MQWS etalon is not very flat, presumbly as a results of the chemical etching and mechanical strains.

We plan to demonstrate optical bistability using MQWS waveguide in the reflected light output. Using MBE, a few um AlAs is grown on the GaAs substrate and then 0.5 um MQWS is grown on the AlAs layer. By using holography techniques, a grating is made on the MQWS surface for exciting a TEoo field in the MQWS waveguide. The wavelength of incident light is close to exciton resonant peak. Due to large nonlinearity of MQWS, the intensity of the reflected light from the grating may show bistable characteristics when the intensity of the incident light changes. Such a device may be advantageous to optical integration. We are doing theoretical calculation of this model.

3.Chaos in Optical Bistability

We began the research of optical bistability in 1979 in liquid crystal hybrid optical bistable device[5] and we have studied chaos in optical bistability since 1981.

The main results we have obtained are as following:

1) The bifurcation diagram of optical bistability and the analytical study of a bimodel map.[6-9]

We calculated the bifurcation diagrams. The symmetry of the bifurcation diagram, the sudden change and hystereses phenomena of chaotic regions were studied. We have also observed the split bifurcation phenomena.

We studied in detail the bifurcation diagram of a bimodel map related to a liquid crystal hybrid optical bistable device. The equations of all the dark lines going through the chaotic region and all boundaries of the chaotic bands are obtained. We showed how to determine the parameter value for a given superstable period from the corresponding word in the symbolic dynamics made of four letters.

2) The transient oscillation in bistable region[10,11]

A new type of self-pulsing oscillation with a period t_R (delay time) has been observed in bistable region. Whether its width is variable or not depends on the initial functions. The oscillation lasts indefinitely. In the case of $\tau/t_R=0$ (τ is the relaxation time of the medium), however, the oscillation is only a transient process.

3) Critical phenomena in optical bistability and chaos[12]

We find that the critical slowing down at the edges of bistable region possesses consistency with the divergence of the time duration of intermittency. The critical exponent is 1/2. We also find that, different from the critical points at the bistable region edges, there are consistencies with the period-doubling bifurcation points at the cusp point. The critical exponent is 1.

4) Frustrated instabilities in optical bistability[13,14]

We studied the unstable behavior of a liquid crystal hybrid bistable device with two time delays in the feedback mechanism. We have observed the frustrated instability, frequency-locking and hysteresis on varying the control parameters. Self-similar structure and Farey-fraction parameter sequences in optical bistability with competing interactions are inverstigated.

5) Influence of feedback strength on the oscillation patterns in optical bistability[15]

We studied the influence of feedback strength on the oscillation patterns in OB with two delayed feedback loops. The frequency locking, frequency jump between neighboring mode, bifurcation, and hysteresis phenomena are observed.

6) First-order Freedericksz transition in nematic liquid crystals[16]

An optical-electric feedback can transform the Freedericksz transition in a nematic film from second order to first order. The criterion that indicate whether the Freedericksz is first or second order are obtained. The hysteresis and bistability accompanying the first-order transition are discussed and observed.

7) Laser-heating-induced bistability in nematic liquid crystals[17]

The intrinsic optical bistability due to the laser-heating-induced nematic-isotropic (N-I) phase transition is observed and analysed. The passive optical intensity-limiting and intensity-switching effects are also obtained. The experimental results are in agreement with the theoretical analysis.

REFERENCES

1. Y.P.Huo, G.Z.Yang, and B.Y. Gu,
   Acta Phys. Sin.**24**,438(1975),**25**,31(1976)
2. G.Z.Yang, Acta Phys. Sin.**30**,1340(1981)
3. Y.S.Chen et al.Acta Phys.Sin.**29**,1307(1980),**33**,1599(1984)
4. S.H.Zheng et al. Acta Phys.Sin.**35**,529(1986)
5. Hong-jun Zhang, Jian-Hua Dai, Jun-Hui Yang and Cun-Xiu Gao,
   Opt. Commun.**38**,21(1981)
6. Z.Hong-jun, D.Jian-hua, W. Peng-ye, in "Laser Spectroscopy
   VI", Springer Ser.Opt.Sci. **40**,322(1983)
7. Zhang Hong-jun, Dai Jian-hua, Wang Peng-ye, Jin Chao-ding,
   and L.M.Narducci,
   Acta Physica Sinica, **33**,1024(1984)
8. Hong-jun Zhang,Jian-hua Dai,Peng-ye Wang,and Chao-ding Jin,
   JOSA B3,231(1986)
9. Dai Jian-hua, Zhang Hong-jun, Wang Peng-ye, Jin Chao-ding
   Acta Physica Sinica, **34**,992(1985)
   Chin. Phys. **6**,14(1986)
   Opt.Commun.**57**,207(1986)
   Zhang Hong-jun,Dai Jian-hua,Wang Peng-ye,Jin Chao-ding,and
   Hao Bai-lin, Chinese Phys. Lett. 2, 5(1985)
10. Wang Peng-ye, Zhang Hong-jun, Dai Jian-hua
   Acta Physica Sinica, **34**,581(1985)
11. Zhang Hong-jun, Dai Jian-hua, Wang Peng-ye,
   Chinese Phys. Lett. 2,129(1985)
12. Wang Peng-ye, Zhang Hong-Jun, Dai Jian-hua,
   Acta Physica Sinica **34**,1223(1985)
   Chin.Phys.6,336(1986)
13. Hong-jun Zhang, and Jian-hua Dai,
   Opt. Lett. **11**,245(1986)
14. Hong-jun Zhang, and Jian-hua Dai,
   IQEC XIV (1986) (invited paper)
15. Jian-hua Dai, and Hong-jun Zhang,
   IQEC XV (1987) (invited paper)
   (IQEC=International Quantum Electronics Conference)
16. Peng-ye Wang, Hong-jun Zhang, and Jian-hua Dai,
   Opt.Lett. **12**, Sep. (1987)
17. Peng-ye Wang, Hong-jun Zhang, and Jian-hua Dai,
   (to be published)

INSTABILITIES IN OPTICAL DELAY DIFFERENTIAL SYSTEMS

By

A. Tallet, M. Le Berre and E. Ressayre

Laboratoire de Photophysique Moléculaire du C.N.R.S. - Bât. 213
Université Paris-Sud
91405 - ORSAY-Cedex, France

ABSTRACT

Instabilities which occur in optical delay-differential systems are numerically investigated. Theoretical analysis of both low-dimension bifurcations and chaos are given.

The dynamics of a passive nonlinear ring cavity and of an electo-optic device are governed by delay-differential (dd) equations [1]. The voltage x(t) of the hybrid device obeys indeed the dd equation

$$\gamma^{-1} \frac{dx}{dt} = - x + f\ [\mu x (t - \tau_R)] \qquad (1)$$

where $f(\mu x)$ is a nonlinear feedback with strength μ and delay τ_R: the dissipative time is δ^{-1}.

Though the phase space has an infinite number of degrees of freedom, the chaotic solutions are shown to evolve in a reduced space with a finite number of independent coordinates, because of the dissipation effects.

In spite of their actual complexity, the chaotic attractors of delay-differential systems have a fractal dimension, the Lyapunov one, d_L, which obeys a quite simple relation :

$$d_L \simeq \frac{\tau_R}{\delta} \qquad , \qquad (2)$$

where δ is the auto-correlation time of the feedback. This law, numerically verified for both optical systems and the Mackey and Glass equation,[3] was originally predicted in the hybrid [4].

The hybrid device obeys Eq. (1) with a periodic feedback, $f(\mu x) = 1 - R \sin \pi \mu x$. In the chaotic regime ($\mu \gg 1$), each time that x increases by a small amount $2\ \mu^{-1}$, the feedback realizes a complete oscillation ; its time trace looks a succession of uncorrelated impulses of short duration δ, that is the mean time required for x(t) to increase by an amount $2\ \mu^{-1}$. Therefore the feedback can be roughly compared to a ramdon force with a finite memory time δ. It follows that the system undergoes τ_R/δ independent excitations during a round-trip time τ_R and has got τ_R/δ independent directions instead of an infinity as it would be with a white noise.

The very intuitive relationship (2) which links the dimension of the attractor with the memory time of the feedback was verified in the hybrid for a large range of τ_R, μ and R (μ R $\gg$ 1). The Lyapunov dimension was deduced from the numerical calculation of the Lyapunov exponents by using the algorithm developed by Farmer [4] and the memory time δ was determined with the help of the correlation function of the feedback.

The conjecture (2) is also realized for both the Mackey-Glass system and the ring cavity device, which do not exhibit any Gaussian character. In the latter, saturation effects inhibit the Gaussian character. When the input intensity increases too much, chaos or more generally any instability vanishes and the system becomes again stable.

The analysis of the route to chaos [5] in a ring cavity is much more complex than the analysis of the chaos itself. There are four relevant control parameters, the input intensity I_0, the absorption coefficient, αl, the initial phase $\Delta\chi$ and the product $\gamma\tau_R$. Generally, αl and $\Delta\chi$ are fixed and I_0 slowly varies along the multistability curve for differents delays.

Roughly speaking, instabilities appear in dispersive media ($\alpha l \ll 1$) with small $\Delta\chi$ around π. Conversely absorptive media ($\alpha l \gg 1$) require large initial phase of magnitude of order several times π for exhibiting instabilities. For given αl, $\Delta\chi$ and I_0, the criticality increases as $\gamma\tau_R$ does.

Until now we have found first Hopf bifurcations with various periods, either near $2\tau_R$ oscillation or near $2\tau_R$-harmonic oscillation, depending on the control parameters, second Hopf bifurcation i.e. a quasi-periodic motion. The ring cavity can also exhibit a saddle-loop bifurcation, corresponding to the first stage of the jump on the uppermost stable branch. Short period-doubling cascade has also been found, followed either by the inverse cascade or by chaos. Another route to chaos has been also identified as a type I intermittency [6].

All these cases have been analyzed with the help of the <u>separation vectors</u>, which are the eigenvectors of the Jacobian operator associated with the dd equations. For these dd systems, the time interval τ_R is divided in N small intervals so that the phase space dimension is approximated by the large number N. Then, solving dd equations amounts to a N dimensional mapping : there are N separation vectors $\Delta\vec{X}_j(k\ \tau_R)$, $j = 1,\ldots,N$, defined on the kth interval τ_R and with nN coordinates, where n is the number of variables (n is equal to unity for the hybrid described by the single scalar quantity x, n is equal to three for the ring cavity described with the help of both a complex field E(t) and a phase $\Phi(t)$). Because of the ambivalence of the role of time, either like a continuous variable or a set of independent coordinates, we can also define separation scalar functions $\Delta Y_j(t)$, ($t = 0, \infty$) which represent the continuous time behavior of the $\Delta\vec{X}_j(n\tau_R)$. In a periodic regime, the Lyapunov exponents $\{\lambda_j\}$ are the real parts of the arguments of the Floquet multipliers divided by the fundamental period T_1,

$$\Lambda_j = e^{(\lambda_j + 2i\pi/T_j)T_1} \qquad (3)$$

The highest Floquet exponent is symbolized by a point on the unit circle ($\Lambda_1 = e^{(\lambda_1 + 2i\pi/T_1)/T_1} = 1$), and any other is inside with coordinates defined by Eq. (3). Any $\Delta Y_j(t)$ is related to the periodic Floquet eigenvector $F_j(t)$ of period T_1, such as

$$\Delta Y_j(t) = e^{(\lambda_j + 2i\Pi/T_j)\,t}\ F_j(t) \qquad (4)$$

with

$$F_j\ (t) = F_j\ (t + T_1) \qquad (5)$$

First of all the time trace of ΔY_1 (t) exhibits the period T_1 of the limit cycle. Second the time trace of any $\Delta Y_j(t)$ which has got a negative λ_j, exhibits its own periodicity T_j. Therefore the alone examination of the time behavior leads to the knowledge of the Floquet matrix. Then varying the input intensity, we directly associate an angle $2\pi T_1/T_j$ to the j-th negative Lyapunov exponent, and we can investigate the respective roles of the Floquet multipliers.

This provides an easy way to analyse the destabilization of the limit cycle and to determine the nature of the bifurcation.

Type I intermittency in a high absorptive ring cavity ($\alpha l = 4$, $\Delta\chi = 6\pi$, $R = 0.9$, $\gamma\tau_R = 10$) was confirmed by the study of the Floquet multipliers below the threshold, $E_c = 1.07394$, as the parameter $\epsilon = E - E_c$ increases from negative values to zero. Far below the threshold $\epsilon \simeq -4\ 10^{-3}$. the nearest Floquet multipliers to the unit circle are complex conjugate with an angle $2\pi T_1/T_2$. of magnitude of order $4\pi/5$. The next Floquet multiplier is real ($\lambda_4 < \lambda_{2,3} < 0$). As ϵ increases ($\epsilon \simeq -3 10^{-4}$ the real Floquet multiplier inside the unit circle overtakes the complex conjugate ones ($\lambda_{2,3} < \lambda_4 < 0$) and finally crosses the unit circle at + 1 as ϵ vanishes.

A return map of this intermittency process displays the role of both the real multiplier, directly responsible for the chaotic bursts and the complex conjugate ones, responsible for the relaminarization.

For quasiperiodic regimes the Floquet theory does not apply and knowledge of the two distinct fundamental frequencies generally results of juggling with the power spectrum lines. The analysis of the separation functions $\Delta Y_j(t)$ is still powerful for quasiperiodic behaviors. It provides an original and reliable way to find out the fundamental frequencies f_1 and f_2. and to analyse the spectrum.

Instabilities in a nonlinear passive ring cavity have been understood in the framework of the linear stability analysis with the help of the Lyapunov vectors and recognized as bifurcations occuring in low dimensional systems. Nevertheless some features have received no explanation within the linear stability analysis and would require a nonlinear one.

REFERENCES

1a. Ikeda, K., Opt. Commun 30, 257 (1979)

1b. Gibbs, H.M., Hopf, F.A., Kaplan, D.L. and Shoemaker, R.L., Phys. Rev. Lett. 46, 474 (1981)

2. Le Berre, M., Ressayre, E., Tallet, A., Gibbs H.M., Kaplan, D.L., Rose, M.H., Phys. Rev. A35, 4020 (1987)

3. Dorizzi, B., Grammaticos, B., Le Berre, M., Ressayre, E., Pomeau, Y., Tallet, A., Phys. Rev. A35 328 (1987).

4. Farmer, J.D., Physica 4D, 366 (1982)

5. Le Berre, M., Ressayre, E., and Tallet, A., in preparation

6. Le Berre, M., Ressayre, E., and Tallet, A., "Type I Intermittency route to chaos in a saturable ring cavity," International workshop on Instabilities, Dynamics and Chaos in nonlinear Optical Systems, Il Ciocco, Italy, July 1987, and in preparation for the Special Issue of J. Opt. Soc. B on "Nonlinear dynamic of lasers", to appear in January 1988.

SELF-PULSING IN LASERS

H.Risken

Abteilung für Theoretische Physik
Universität Ulm
D-7900 Ulm, West Germany

1 INTRODUCTION

The ideal laser emits the laser light in form of a continuous wave (cw). This can of course only happen if the pumping of the active atoms is done continuously, if any other parameters do not change in time and if we neglect spontaneous emission noise. By suddenly varying the cavity losses (Q-switching) or some other parameters of the laser system one may obtain frequency locking of the modes which gives rise to an emission of the laser light in form of a pulse train (forced pulsing). The emission of laser light in form of pulses, however, can also occur, if all the parameters of the laser system are kept constant in time. In such a case we talk of self-pulsing. It is the main purpose of the present paper to show for a simple laser model that for certain parameters the laser equations lead to such self-pulsing behavior. Though the laser equations admit cw solutions, such cw solutions may become unstable for certain parameter values and may thus be no longer observable. Thus the investigation of the laser instabilities is of major importance for self-pulsing.

The instabilities in lasers change mainly with increasing pumping strength. The onset of laser action for one mode is the first instability. By increasing further the pumping strength new instabilities occur. Depending on the parameters and on the model, either the intensity of the one mode will start to oscillate or some off-resonance modes will become unstable. The last instability then leads to a build up of stable light pulses, which travel through the laser cavity, i.e.

we observe self-pulsing. If the pumping strength is increased further, the light pulses themselves are no longer stable. They change their height and their shape in a periodic or even chaotic way. This effect may be termed periodic or chaotic breathing of the pulses. It should be noted that the one-mode laser equations are equivalent to the Lorenz equations. Therefore for appropriate parameters the one-mode laser equations will also show chaotic behaviour.

2 MODEL AND BASIC EQUATIONS

In our investigations we want to restrict ourselves to the simplest case. We therefore assume that the active material consists of two-level atoms with a homogeneously broadened line. In addition we assume that the cavity losses are homogeneously distributed in the cavity and that we have only one direction of polarization and that the field depends only on the time and on the coordinate x of the laser axis. The **basic equations** [1,2,3] are: 1) the one-dimensional wave equation for the electric field $\tilde{\tilde{E}}(x,t)$ (telegrapher's equation) driven by the polarization $\tilde{\tilde{P}}(x,t)$, i.e.

$$\frac{\partial^2 \tilde{\tilde{E}}}{\partial x^2} - \frac{1}{c^2}\frac{\partial^2 \tilde{\tilde{E}}}{\partial t^2} - \frac{2\kappa}{c^2}\frac{\partial \tilde{\tilde{E}}}{\partial t} = \mu \frac{\partial^2 \tilde{\tilde{P}}}{\partial t^2}, \tag{2.1}$$

2) the material equations for polarization $\tilde{\tilde{P}}(x,t)$ and inversion $\tilde{\tilde{S}}(x,t)$ driven by the electric field, i.e.

$$\frac{\partial^2 \tilde{\tilde{P}}}{\partial t^2} + 2\gamma_2 \frac{\partial \tilde{\tilde{P}}}{\partial t} + \omega_0{}^2 \tilde{\tilde{P}} = -2(M^2/\hbar)\omega_0 \tilde{\tilde{E}}\tilde{\tilde{S}}$$

$$\frac{\partial \tilde{\tilde{S}}}{\partial t} = \gamma_1(d_0 - \tilde{\tilde{S}}) + \frac{2}{\hbar\omega_0}\frac{\partial \tilde{\tilde{P}}}{\partial t}\tilde{\tilde{E}} \tag{2.2}$$

and 3) appropriate boundary conditions. In (2.1,2) κ describes the losses, μ is the permeability, $c = 1/\sqrt{\varepsilon\mu}$ the velocity of light in the host material, M the dipole matrix element and $\hbar\omega_0$ the energy separation of the two levels of the active atoms. The constants γ_1 and γ_2 describe the damping of the inversion and polarization respectively and d_0 describes the strength of the pumping. Here we want to consider a ring-laser system, see Fig. 1. Therefore we have the boundary conditions

$$\tilde{\tilde{E}}(x+L,t) = \tilde{\tilde{E}}(x,t), \quad \tilde{\tilde{P}}(x+L,t) = \tilde{\tilde{P}}(x,t), \quad \tilde{\tilde{S}}(x+L,t) = \tilde{\tilde{S}}(x,t) \ , \quad (2.3)$$

where L is the cavity length.

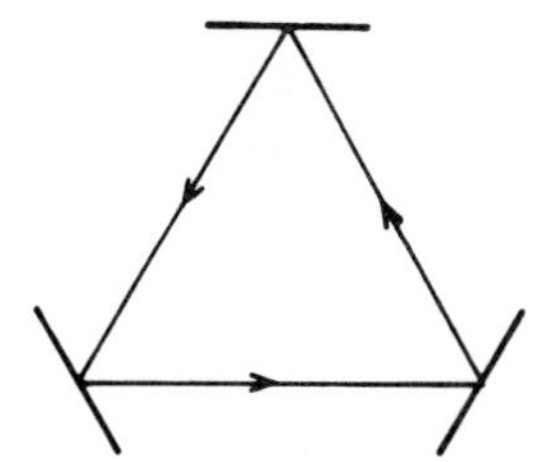

Fig. 1. Ring-laser cavity

Because the frequency ω_0 is usually much larger than any other time constant involved, (2.1) and (2.2) can be simplified by the **rotating wave approximation.** Assuming only one direction of propagation we write

$$\begin{aligned} \tilde{\tilde{E}} &= \tilde{E} \exp[i\omega_0(x/c-t)] + \text{c.c.} \ , \\ \tilde{\tilde{P}} &= \tilde{P} \exp[i\omega_0(x/c-t)] + \text{c.c.} \ , \\ \tilde{\tilde{S}} &= \tilde{S} \ , \end{aligned} \qquad (2.4)$$

where $\tilde{E}(x,t)$, $\tilde{P}(x,t)$ are slowly varying complex envelopes in x and t and where + c.c means that we have to add the complex conjugate. If higher harmonics are neglected the inversion $\tilde{S}$ will also slowly vary in x and t. Neglecting higher order harmonics the insertion of (2.4) into (2.1,2) leads to ($\partial/\partial t$ is now denoted by a dot)

$$\begin{aligned} c(\partial\tilde{E}/\partial x) + \dot{\tilde{E}} + \kappa\tilde{E} &= i(\omega_0/2\varepsilon)\tilde{P} \\ \dot{\tilde{P}} + \gamma_2\tilde{P} &= -i(M^2/\hbar)\tilde{E}\tilde{S} \end{aligned}$$

$$\dot{\tilde{S}} + \gamma_1\tilde{S} = \gamma_1 d_0 + (2i/\hbar)(\tilde{E}\tilde{P}^* - \tilde{E}^*\tilde{P}) \ . \qquad (2.5)$$

If we assume that **one cavity mode is in resonance** with the atomic frequency ω_0, i.e. $\exp(i\omega_0 L/c)=1$, the boundary conditions for the slowly varying fields $\tilde{E}$, $\tilde{P}$, $\tilde{S}$ will be the same as those in (2.3) for the rapidly varying fields.

For $d_0<\tilde{S}_{cw}$ where $\tilde{S}_{cw}$ is defined in (2.6) we have only the trivial solution $\tilde{P}=\tilde{E}=0$, $\tilde{S}=d_0$. This trivial solution is stable for $d_0<\tilde{S}_{cw}$ but unstable for $d_0>\tilde{S}_{cw}$.

The equations (2.5) have a nontrivial solution for d_0 larger than $\tilde{S}_{cw}$, which is independent of x and t. This cw solution is given by

$$\tilde{S}_{cw} = (2\varepsilon\hbar\kappa\gamma_2)/(\omega_0 M^2)$$

$$\tilde{E}_{cw} = \sqrt{(d_0-\tilde{S}_{cw})\gamma_1\hbar\omega_0/(2\kappa\varepsilon)/2}$$

$$\tilde{P}_{cw} = -i(2\varepsilon\kappa/\omega_0)\tilde{E}_{cw} \ . \qquad (2.6)$$

It will be useful to introduce the following **normalized fields**

$$E = \tilde{E}/\tilde{E}_{cw}, \quad P = \tilde{P}/\tilde{P}_{cw}, \quad \sigma = \tilde{S}/\tilde{S}_{cw}. \qquad (2.7)$$

The equations (2.5) then reduce to the following normalized form [4]

$$\begin{aligned} &c(\partial E/\partial x) + \dot{E} + \kappa E = \kappa P \\ &\dot{P} + \gamma_2 P = \gamma_2 E\sigma \\ &\dot{\sigma} + \gamma_1\sigma = \gamma_1[\lambda + 1 - \lambda(E^*P + EP^*)/2] \ , \end{aligned} \qquad (2.8)$$

where the normalized pump parameter λ is given by

$$\lambda = (d_0 - \tilde{S}_{cw})/\tilde{S}_{cw} \ . \qquad (2.9)$$

3 THRESHOLD INSTABILITY

If we consider only the particular cavity mode, which is in resonance with the atomic frequency ω_0, then the slowly varying fields will not depend on x. Furthermore, if κ is much smaller than the other damping constants γ_1 and γ_2 and if we are near threshold (small $|d_0 - \tilde{S}_{cw}|$), we can neglect the time derivatives in the last two equations in (2.5). Eliminating $\tilde{P}$ and $\tilde{S}$ we thus arrive at

$$\dot{\tilde{E}} + [\kappa - \frac{\omega_0 M^2}{2\varepsilon\hbar\gamma_2} \frac{d_0}{1+4M^2\tilde{E}^*\tilde{E}/(\hbar^2\gamma_1\gamma_2)}]\tilde{E} = 0 \ . \quad (3.1)$$

Near threshold the intensity $\tilde{E}^*\tilde{E}$ is a small quantity and we may therefore expand the denominator which then yields

$$\dot{\tilde{E}} = \bar{\delta}(d - \tilde{E}^*\tilde{E})\tilde{E} \quad (3.2)$$

with

$$\bar{\delta}d = (\omega_0 M^2 d_0)/(2\varepsilon\hbar\gamma_2) - \kappa \ ; \ \bar{\delta} = (2\omega_0 M^4 d_0)/(\varepsilon\gamma_2{}^2\hbar^3\gamma_1) > 0 \ . \quad (3.3)$$

Eq. (3.2) is the rotating-wave Van der Pol equation. The Van der Pol equation

$$\ddot{y} - 2\bar{\delta}(d - y^2)\dot{y} + \omega_0{}^2 y = 0 \quad (3.4)$$

is one of the simplest equations, which describe a self-sustained oscillator with frequency ω_0. Because in (3.2) the phase of $\tilde{E}$ does not change we may as well use the intensity $\tilde{I}=\tilde{E}^*\tilde{E}$. We then obtain from (3.2) the Verhulst equation

$$\dot{\tilde{I}} = 2\bar{\delta}(d - \tilde{I})\tilde{I} \ , \quad (3.5)$$

which is very well known in population dynamics, see for instance [5a]. The stationary solutions of (3.5) are obviously given by

$$\tilde{I}_{st}^{(1)} = 0 \qquad \text{for } d \leq 0 \text{ (at and below threshold)}$$

$$\left.\begin{aligned} \tilde{I}_{st}^{(1)} &= 0 \\ \tilde{I}_{st}^{(2)} &= d\,. \end{aligned}\right\} \quad \text{for } d > 0 \text{ (above threshold)} \tag{3.6}$$

(Because the intensity $\tilde{I}$ must be positive, the solution $\tilde{I}_{st}^{(2)}=d$ is physically irrelevant for $d<0$.) Though the differential equation (3.5) can be solved exactly we shall investigate the stability of the stationary solution for $d\neq 0$ by a **linearization procedure.** This method is also applicable to more complicated systems which appear in Chaps. 4 and 5. First we take into account **small deviations of the stationary solution**

$$\tilde{I} = \tilde{I}_{st} + \Delta\tilde{I}\,. \tag{3.7}$$

Next by inserting (3.7) into (3.5) and retaining only the linear terms in $\Delta\tilde{I}$ we get the following linear differential equations

$$\Delta\dot{\tilde{I}}^{(1)} = -2\bar{\delta}|d|\Delta\tilde{I}^{(1)} \qquad d < 0$$

$$\left.\begin{aligned} \Delta\dot{\tilde{I}}^{(1)} &= 2\bar{\delta}d\Delta\tilde{I}^{(1)} \\ \Delta\dot{\tilde{I}}^{(2)} &= -2\bar{\delta}d\Delta\tilde{I}^{(2)} \end{aligned}\right\} \quad d > 0 \qquad . \tag{3.8}$$

Finally the ansatz

$$\Delta\tilde{I}(t) = \Delta\tilde{I}_0\cdot\exp(\beta t) \tag{3.9}$$

leads to

$$\beta^{(1)} = -2\bar{\delta}|d| \qquad d < 0$$

$$\left.\begin{aligned} \beta^{(1)} &= 2\bar{\delta}d \\ \beta^{(2)} &= -2\bar{\delta}d \end{aligned}\right\} \quad d > 0 \tag{3.10}$$

Obviously the solution becomes unstable if β is larger than zero, it is stable if β is less than zero. If β is equal to zero higher order terms in $\Delta\tilde{I}$ must be taken into account. Thus the stationary solution $\tilde{I}_{st}^{(1)}=0$ is stable below threshold but unstable above threshold. The other solution $\tilde{I}_{st}^{(2)}=d$, which only exists above threshold, is stable, see Fig. 2. (At threshold d=0, β is equal to zero. As seen from (3.5) the solution $\tilde{I}_{st}^{(1)}=0$ is still stable for d=0.)

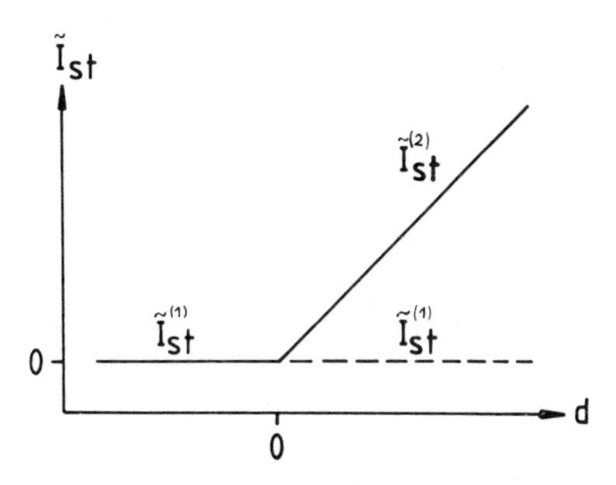

Fig. 2. Stable (solid line) and unstable (broken line) stationary solutions of (3.5)

4 SINGLE-MODE INSTABILITY

Now we want to investigate the stability of the cw solution (2.6) for the resonant mode. Therefore the field envelope does not depend on x. Furthermore we consider only real values of E and P (i.e. we assume that the phases as given by (2.6) are constant). Eqs. (2.8) then reduce to

$$\begin{aligned}\dot{E} + \kappa E &= \kappa P \\ \dot{P} + \gamma_2 P &= \gamma_2 E\sigma \\ \dot{\sigma} + \gamma_1 \sigma &= \gamma_1(\lambda + 1 - \lambda EP) \ .\end{aligned} \tag{4.1}$$

The main trick in investigating the stability of a solution of nonlinear equations like (3.1) is to introduce additional small deviations to the known solution. Because of their smallness one can neglect products of the small deviations and thus obtain linear equations which are much easier to solve than the original nonlinear ones. Introducing the small deviations e, p, δ from the cw solution ($E = P = \sigma = 1$), the **linearized equations** for e, p, δ then read

$$\dot{e} + \kappa(e - p) = 0$$
$$\dot{p} + \gamma_2(p - e - \delta) = 0$$
$$\dot{\delta} + \gamma_1[\delta + \lambda(e+p)] = 0 . \qquad (4.2)$$

The ansatz

$$\begin{pmatrix} e \\ p \\ \delta \end{pmatrix} = \begin{pmatrix} e_0 \\ p_0 \\ \delta_0 \end{pmatrix} \cdot \exp(\beta t) \qquad (4.3)$$

then leads to the cubic equation

$$\beta^3 + (\gamma_1+\gamma_2+\kappa)\beta^2 + \gamma_1(\gamma_2+\kappa+\gamma_2\lambda)\beta + 2\gamma_1\gamma_2\kappa\lambda = 0 . \qquad (4.4)$$

It follows from the Hurwitz criterion that the solutions (4.3) are stable, i.e. have a negative real part, if

either $\gamma_1 + \gamma_2 \geqq \kappa$

or $\gamma_1 + \gamma_2 < \kappa; \quad \lambda \leqq \lambda_s$. (4.5)

Here λ_s is given by [5]

$$\lambda_s = (\gamma_1+\gamma_2+\kappa)(\gamma_2+\kappa)/[(\kappa-\gamma_1-\gamma_2)\gamma_2] . \qquad (4.6)$$

The solutions are unstable if

$$\gamma_1 + \gamma_2 < \kappa, \qquad \lambda > \lambda_s . \tag{4.7}$$

Thus in order to get unstable solutions for the one-mode case the losses of the cavity must exceed the sum of the losses of the polarization and the inversion (**bad cavity case**). In the **good cavity case** $\kappa<\gamma_1+\gamma_2$, the cw one-mode solution is always stable.

As it was pointed out by HAKEN [3,6] Eqs. (4.1) are equivalent to the Lorenz equations. Introducing

$$\tau = \gamma_2 t; \quad X = \sqrt{\lambda\gamma_1/\gamma_2}E; \quad Y = \sqrt{\lambda\gamma_1/\gamma_2}P;$$
$$Z = \lambda + 1 - \sigma \tag{4.8}$$

(4.1) are transformed to

$$dX/d\tau = a(Y - X)$$
$$dY/d\tau = X(r - Z) - Y$$
$$dZ/d\tau = XY - bZ , \tag{4.9}$$

which are identical to the Lorenz equations [7]. Here a, b, r are given by

$$a = \kappa/\gamma_2, \quad b = \gamma_1/\gamma_2, \quad r = \lambda + 1 .$$

Therefore, in addition to a limit cycle solution, chaotic solutions are also possible. It should be noted that the condition (4.6,7) (in the notation of the Lorenz equation (4.9)) has been derived by Lorenz himself [7].

5 INSTABILITY OF NONRESONANT MODES

In the good cavity case a single resonant mode is always stable. In the presence of the resonant mode other nonresonant modes, however, may become unstable. To investigate this instability we again introduce **small deviations** e, p, σ, ϕ, ψ from the cw solutions, i.e.

$$E = 1 + e \cdot \exp(i\phi); \quad P = 1 + p \cdot \exp(i\psi); \quad \sigma = 1 + \delta \,. \tag{5.1}$$

The nonlinear system (2.8) thus reduces to the linear system

$$\begin{aligned}
&c(\partial e/\partial x) + \dot{e} + \kappa(e-p) = 0 \\
&\dot{p} + \gamma_2(p-e-\delta) = 0 \\
&\dot{\delta} + \gamma_1[\delta + \lambda(e+p)] = 0 \\
&c(\partial \phi/\partial x) + \dot{\phi} + \kappa(\phi-\psi) = 0 \\
&\dot{\psi} + \gamma_2(\psi-\phi) = 0 \,.
\end{aligned} \tag{5.2}$$

The ansatz

$$\begin{pmatrix} e \\ p \\ \delta \\ \phi \\ \psi \end{pmatrix} = \begin{pmatrix} e_0 \\ p_0 \\ \delta_0 \\ \phi_0 \\ \psi_0 \end{pmatrix} \cdot \exp[i(\alpha/c)x + \beta t] + \text{c.c.} \tag{5.3}$$

leads to the characteristic equations

$$\begin{aligned}
\beta^3 + (\gamma_1+\gamma_2+\kappa+i\alpha)\beta^2 + [\gamma_1(\gamma_2+\kappa+\gamma_2\lambda)+(\gamma_1+\gamma_2)i\alpha]\beta& \\
+ 2\gamma_1\gamma_2\kappa\lambda + \gamma_1\gamma_2(1+\lambda)i\alpha = 0&
\end{aligned} \tag{5.4}$$

$$\beta^2 + (\kappa+\gamma_2+i\alpha)\beta + i\gamma_2\alpha = 0 \,. \tag{5.5}$$

Because of the boundary condition (2.3) the wave number α/c should have the values

$$\alpha/c = m \cdot 2\pi/L; \quad m=0,\ \pm1,\ \pm2,\ \ldots\ . \tag{5.6}$$

The values for $m \neq 0$ correspond to the different off-resonant modes. It can be derived from (5.4) that the real part of β can become positive if the pump parameter exceeds a critical value [4,8,9]

$$\lambda > \lambda_c = 4 + 3(\gamma_1/\gamma_2) + 2\sqrt{4 + 6(\gamma_1/\gamma_2) + 2(\gamma_1/\gamma_2)^2}\ . \tag{5.7}$$

The dispersion relations $\beta(\alpha)$ for two typical cases with $\lambda > \lambda_c$ are shown in Figs. 3, 4.

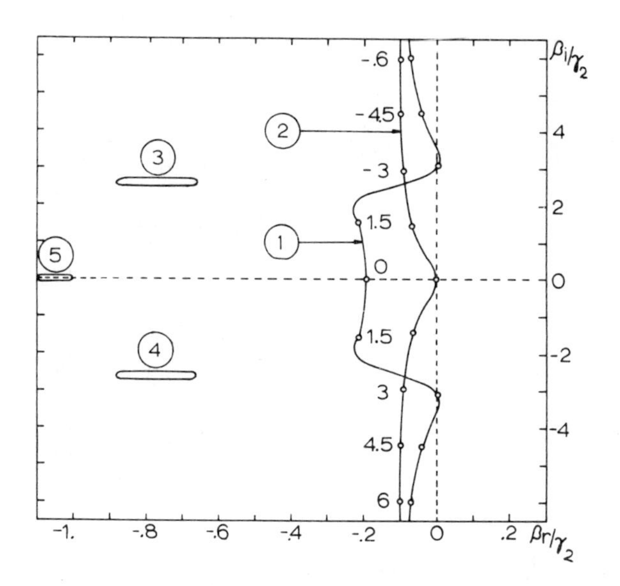

Fig. 3. Roots of the characteristic equations (5.4) and (5.5) with α varying from $-\infty$ to $+\infty$. The fixed parameters are $\gamma_1 = \gamma_2/2$, $\kappa = \gamma_2/10$ and $\lambda = 15$. (1), (3) and (4) are the roots corresponding to the amplitudes, (2) and (5) are the roots corresponding to the phases. The instable region of (1) is $2.9\gamma_2 < |\alpha| < 3.6\gamma_2$. The discrete values α_n are shown for $2\pi c/L = 1.5\gamma_2$ for root loci (1) and (2). Two instable modes are $\alpha_2 = -3\gamma_2$ and $\alpha_{-2} = 3\gamma_2$

Thus if the discrete α values lie in the appropriate region, these off-resonances become unstable. For the lower part on Fig. 3 and 4 the corresponding region for α reads [4]

$$\alpha_{min} < \alpha < \alpha_{max} ;$$

$$\alpha_{\substack{max\\min}} = \sqrt{\gamma_1(3\gamma_2\lambda - \gamma_1 \pm R)/2}\cdot\{1 - 2\kappa/[\gamma_2(\lambda-2)-\gamma_1 \pm R]\} ;$$

$$R = \sqrt{\lambda^2\gamma_2{}^2 - 2(3\gamma_1+4\gamma_2)\lambda\gamma_2 + \gamma_1{}^2} \quad . \tag{5.8}$$

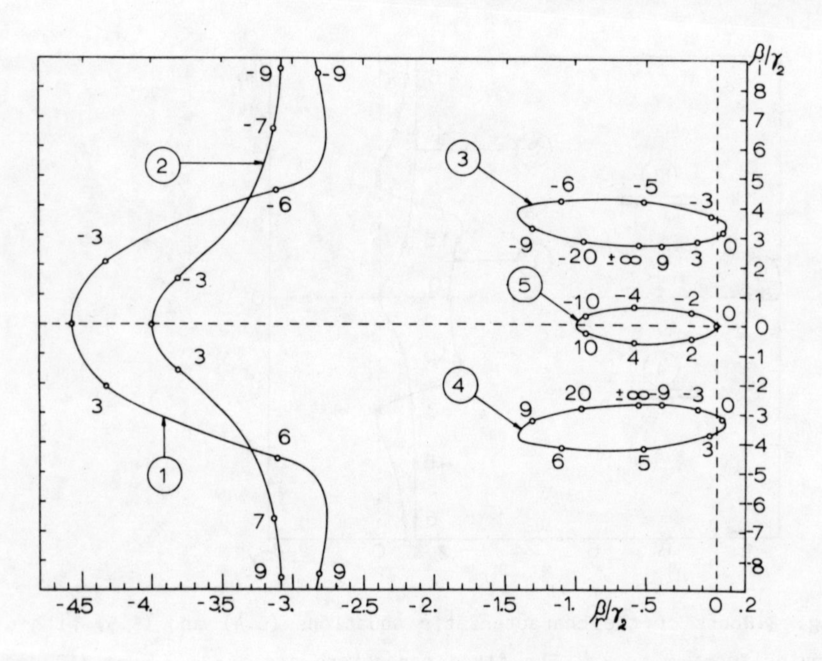

Fig. 4. Roots of the characteristic equations (5.4) and (5.5) with α varying from $-\infty$ to $+\infty$ as marked on the curves. The fixed parameters are $\gamma_1=\gamma_2/2$, $\kappa=3\gamma_2$ and $\lambda=15$. (1), (3) and (4) are the roots corresponding to the amplitudes, (2) and (5) are the roots corresponding to the phases. The instable regions of (3) and (4) are $-2.5\gamma_2<\alpha<0.9\gamma_2$ and $-0.9\gamma_2<\alpha<2.5\gamma_2$, respectively

One remarkable feature of (5.7) is that it is independent of κ. It can be shown that the critical value (4.6) for varying $\kappa>\gamma_1+\gamma_2$ is larger than or equal to the above critical value ($\lambda_s \geqq \lambda_c$). The minimum $\lambda_s=\lambda_c$ occurs for $\kappa=\kappa_{min}= \gamma_1+\gamma_2[1+\sqrt{4+6(\gamma_1/\gamma_2)+2(\gamma_1/\gamma_2)^2}]$. Therefore λ_c is also a lower bound of λ for the instability of a single mode in the bad cavity case. (It may be shown that for $\lambda>\lambda_s \geqq \lambda_c$ α_{min} becomes negative in agreement with the instability results of the previous chapter for a single mode, i.e. for $\alpha=0$.)

In addition to the tuned cw solution $E=P=\sigma=1$, detuned cw solutions of the system (2.8) are also possible. These solutions are given by ($\Omega^2<\lambda$) [11]

$$\begin{aligned} E &= \sqrt{1-\Omega^2/\lambda}\cdot\exp\{i\Omega[(\gamma_2+\kappa)x/c-\gamma_2 t]\} \\ P &= E(1 + i\Omega) \\ \sigma &= 1 + \Omega^2 \ . \end{aligned} \tag{5.9}$$

GERBER and BÜTTIKER have shown [11] that the solution (5.9) becomes unstable if the parameter Ω is large enough. Because of $\lambda>\Omega^2$ the pump parameter λ must then also be large. In this case the phase branch in Fig. 3 crosses the imaginary axis.

6 TRANSIENT BUILDUP OF THE PULSE

If one of the nonresonant modes is unstable, the deviations e, p, and δ grow to infinitly large values. Thus the linearized equations (5.2) cannot be used for larger times. Therefore one has to solve the original nonlinear equations (2.8). It can be shown [4] that the phases of E and P in (2.8) are stable in the nonlinear region. Thus only the amplitude equations

$$c\partial|E|/\partial x + |\dot{E}| + \kappa|E| = \kappa|P|$$

$$|\dot{P}| + \gamma_2|P| = \gamma_2|E|\sigma$$
$$\dot{\sigma} + \gamma_1\sigma = \gamma_1(\lambda + 1 - \lambda|E||P|) \qquad (6.1)$$

have to be solved. There are two main methods to integrate these nonlinear equations numerically.

In the first method one uses a discretisation of the variables x and t. If the differential quotients are replaced by appropriate difference quotients one gets difference equations which can then be solved numerically. When using this procedure one must be careful to obtain stable difference equations, i.e. those where the differences of physical quantities belonging to neighboring space points do not blow up, see for instance [10]. In the second method one expands the fields into the modes $\exp(i(\alpha/c)x]$ with α given by (5.6) and solves the system of truncated nonlinear ordinary differential equations for the expansion coefficients numerically.

As shown in [4] by the first method a pulse builds up if one of the nonresonant modes is unstable. The stable explicit difference equations approximating (6.1) correct to second order differences are given in Appendix I of [4]. In Fig. 5 the transient solutions of (6.1) are shown. As initial values we chose a (small) instable solution of the linearized equation. The figure shows how the amplitudes of the waves become saturated and how the waveform of the electric field is changed from a pure sinusiodal form to a more peaked waveform. The maximum and minimum values of the electric field are compared with the values that follow from the linearized equations in Fig. 6. In the same figure, the transient velocity (defined as the velocity of the peak of the electric field amplitude) is also plotted as a function of the number of round trips N. Notice that the pulse velocity decreases with increasing pulse height, but always remains larger than the velocity c in the host material. The final pulse has the form $f(t-x/v)$ and thus describes a pulse traveling with the velocity v through the cavity.

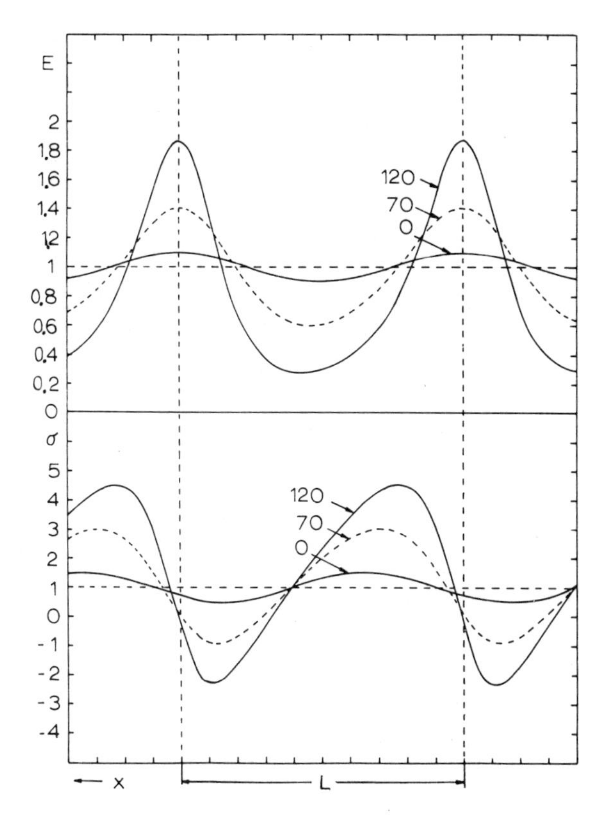

Fig. 5. The transient waveform of the electric field E and inversion σ as a function of x for different times t=NL/c. (N is the number of round trips that light, traveling with the velocity c through the cavity, makes in the time t.) The direction of x is to the left so that the pulse travels to the left, consistent with Fig. 8. The parameters are $\gamma_1=\gamma_2/2$, $\kappa=\gamma_2/10$, $\lambda=15$, and $L=2\pi c/(3.2\gamma_2)$. Thus, α is in the instable region of Fig. 3. As initial conditions we have chosen an instable solution (5.3) with $e_0=0.1$. (The values p_0 and δ_0 then follow from the eigenvector equation.) In the figure, the pulses are shifted in x in such a way that the minimum values always coincide

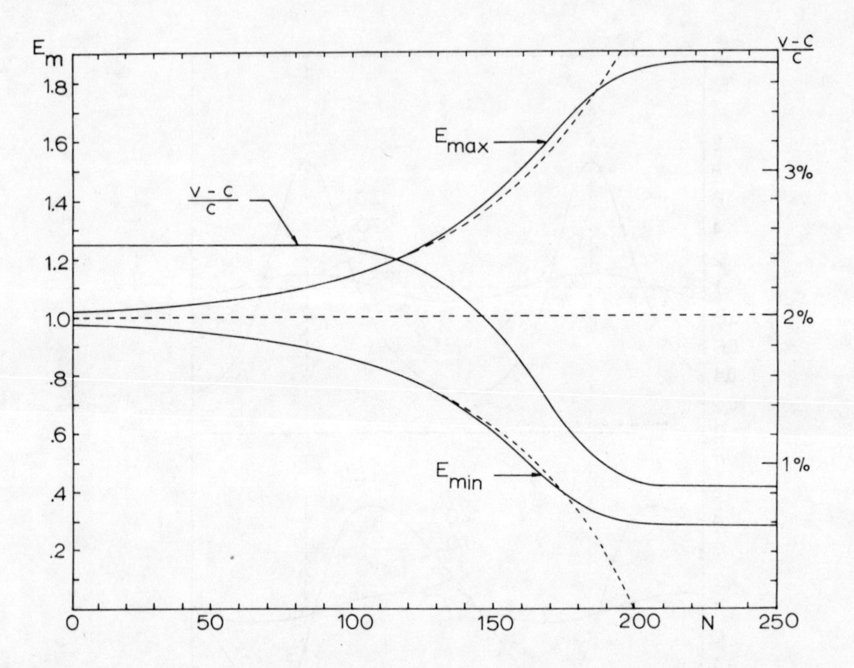

Fig. 6. The maximum and minimum amplitudes of the electric field together with the relative excess velocity (v-c)/c as a function of the time t=NL/c. v is the velocity of the maximum value of the electric field. The parameters are $\gamma_1=\gamma_2/2$, $\kappa=\gamma_2/10$, $\lambda=15$, and $L=2\pi c/(3.2\gamma_2)$. Thus, α is in the instable region of Fig. 3. As initial conditions we have chosen an instable solution (5.3) with $e_0=0.02$. (The values p_0 and δ_0 then follow from the eigenvector equation.) The time variation of E_{max} and E_{min} that follows from the linear theory (5.3-5.5) is dotted in for comparison

Usually the laser oscillation does not start at $|E|\approx|p|\approx\sigma\approx1$. If instead we assume a fast Q switching of the laser resonator, the initial conditions are given by $\sigma=\lambda+1$, $|E|=|P|=0$. Because we have neglected noise in our treatment, the solutions would remain $|E|=|P|=0$ for all times. In order to get the laser started, one must therefore assume a small initial disturbance of the electric field, e.g., a Gaussian. The transient buildup of the pulse from such a small disturbance is shown in Fig. 7. After a few round trips, the variables reach the approximate cw soutions $E \approx P \approx \sigma \approx 1$, but for the parameters of Fig. 7, these values are not stable. Hence, a pulse builds up with the same final shape as in Fig. 5. In the first few round trips the field distribution is rather complicated. After these few round trips, the field variables have the form of a traveling wave pulse $f(t-x/v,t)$, which slowly changes its amplitude, its velocity, and its pulse shape (see Figs. 6 and 5). For similar initial conditions as in Fig. 7, but with a pump parameter λ below the critical value λ_c, the transient solution shows in the beginning a similar behavior as in Fig. 7. But then, after a few round trips, the stationary cw values $|E|=|P|=\sigma=1$ are reached.

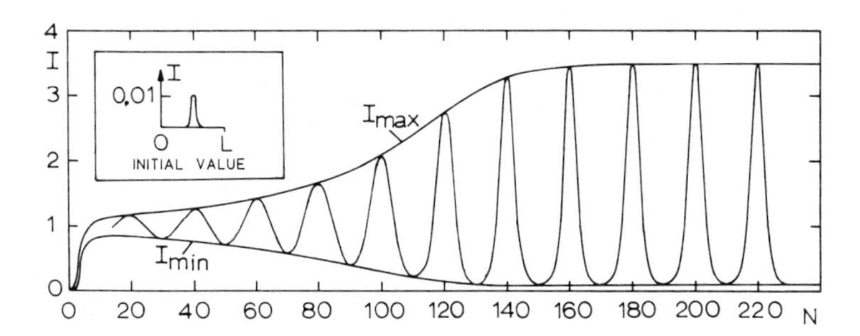

Fig. 7. The transient buildup of the intensity from a small Gaussian disturbance as a function of the time $t=NL/c$ in the instable region $\lambda=15$, $L=2\pi c/(3.2\gamma_2)$, $\gamma_1=\gamma_2/2$, and $\kappa=\gamma_2/10$. The initial values are given by $E(x,0)=0.1 \cdot \exp[-100(x/L-1/2)^2]$; $P(x,0)=0$, $\sigma(x,0)=\lambda+1$. Only every 20th pulse is shown with a width 20 times the actual width

7 THE STEADY-STATE PULSE

The transient solutions show that there exists a steady-state pulse of the form

$$|E| = E(t-x/v); \quad |P| = P(t-x/v); \quad \sigma = \sigma(t-x/v) \tag{7.1}$$

of the equation (6.1), where v is the velocity of the pulse. Putting (7.1) into the partial differential Equations (6.1) leads to a set of ordinary differential equations:

$$\varepsilon E' + E = P \tag{7.2}$$

$$P' + \gamma_2 P = \gamma_2 E\sigma \tag{7.3}$$

$$\sigma' + \gamma_1\sigma = \gamma_1(\lambda + 1 - \lambda EP) \tag{7.4}$$

where the prime means differentiation with respect to t-x/v and where ε is connected with the velocity v through the relations

$$\varepsilon = (1/\kappa)[1 - (c/v)], \quad v/c = 1/(1 - \varepsilon\kappa). \tag{7.5}$$

The periodic boundary conditions (2.3) require the variables E, P, and σ to be periodic in L/v. The solutions of the equations (7.2-7.4) with the boundary conditions (2.3) can be obtained be solving (7.2-7.4) by a Runge-Kutta procedure for one period. The initial values of E, P, and σ and the parameter ε have to be determined such that the boundary conditions (2.3) are fulfilled, see Appendix II of [4] for more details. The values of E, P, and σ for a steady state pulse are shown in Fig. 8. It also seems to be of interest that for certain values of L and for $\lambda>\lambda_c$ both the cw solution and the pulse solution can be stable (bistable operation), see Fig. 9.

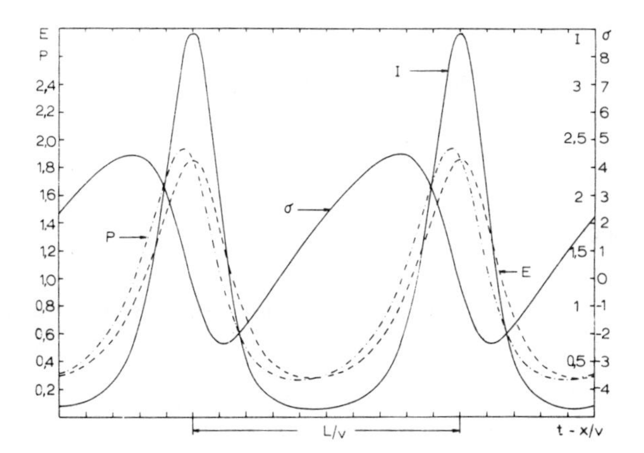

Fig. 8. Steady-state fields E, P (dotted) and σ, $I=E^2$ (solid) vs. t-x/v for $\gamma_1=\gamma_2/2$, $\lambda=15$, $L/v=2\pi/(3.47\gamma_2)$. Direction of propagation is from right to left along the x axis. The one-period average intensity is $\langle I\rangle=0.973$, the average inversion $\langle\sigma\rangle=1.40$

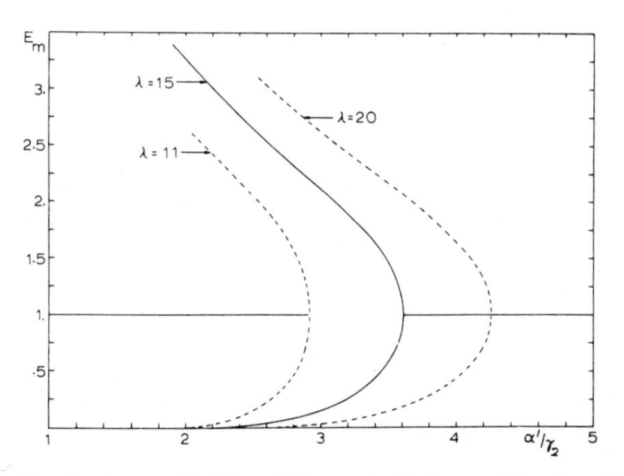

Fig. 9. Maximum ($E_m=E_{max}>1$) and minimum ($E_m=E_{min}<1$) electric field pulse amplitudes as a function of $\alpha'=2\pi v/L$ for $\gamma_1=\gamma_2/2$ and for three pump parameters $\lambda=20$ (dotted line), $\lambda=15$ (solid line), $\lambda=11\approx\lambda_c$ (dotted line); v is the velocity of the corresponding pulse and is approximately equal to the velocity of light in the host material for $\kappa\ll\gamma_2$. The instable region of the cw solution for $\lambda=15$ is marked by a

blank space between $2.92\gamma_2<\alpha'<3.62\gamma_2$. Since for $\alpha'>\alpha_{max}'=\alpha_{max}/(1-\varepsilon\kappa)$ a velocity v is not defined, we may use for $\alpha'\geq\alpha_{max}'$, $v=v(\alpha_{max}')$

8 INSTABILITY OF PULSES

If the pump parameter is increased still further the pulses themselves will change. The maximum value of the pulse, its minimum value and its shape will alter in time while it is traveling through the cavity. This behaviour has been termed breathing of the pulse. Breathing of the pulse can occur in a periodic or in a chaotic manner. The maximum intensity of the pulses as a function of time and its Fourier transform are shown in Figs. 10 and 11 for the case of periodic and chaotic breathing [12].

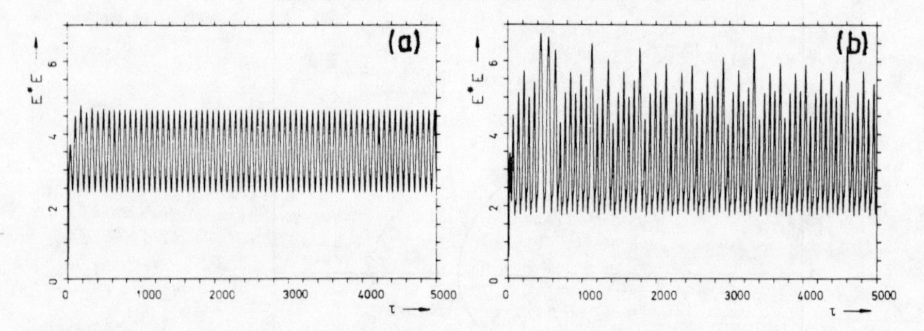

Fig. 10. The maximum intensity E^*E in the cavity as a function of the normalized time $\tau=\gamma_2 t$ as obtained by a numerical integration of the coupled equations for the expansion coefficients. a) Periodic breathing. Parameters $\lambda=20$, $\Omega^2=9$, $\gamma_1=\gamma_2/2$, $\kappa=\gamma_2/10$. b) Chaotic breathing. Parameters $\lambda=25$, $\Omega^2=10$, $\gamma_1=\gamma_2/2$, $\kappa=\gamma_2/10$. The initial values of both cases are the Gerber-Büttiker solutions where the Fourier coefficients of the inversion have been perturbed by adding 0.01 to all coefficients

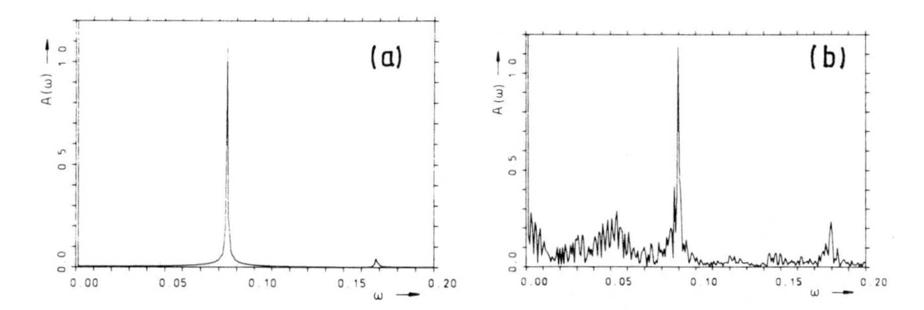

Fig. 11. Fourier spectrum of the maximum intensity corresponding to Fig. 10. The amplitude A(ω) is plotted versus ω

9 CONCLUSION

We have seen that the equations for a simple laser model (ring-laser with one direction of propagation , spatially homogeneously distributed loss and gain, two-level atoms with a homogeneously broadened line, one direction of polarization) show a great variety of instabilities. The simple model treated so far can be extended in several ways. One may in addition to the active material include a passive material [13-17]. The extension to an inhomogeniously broadened line was given by CASPERSON [18]. A paper by LUGIATO et al. [19] treats the case where the loss of the electric field in the cavity occurs at one of the mirrors. Selfpulsing in ring-lasers with bistable absorption was discovered some time ago, see for instance [20,21] for reviews. Furthermore transverse effects seem to play a role in the investigations of instabilities [22]. One should also mention the Ikeda instability [23], where the essential reason for its occurence seems to be a delay mechanism. In an experiment done with a ring-laser system having a homogeneously broadened line a number of instabilities have been observed [24,25].

REFERENCES

1 H.Haken, Laser Theory, in Encyclopedia of Physics, Vol. XXV/2c, Springer, Berlin (1970)

2 M.Sargent, M.O. Scully, W.E. Lamb, Laser Physics, Addison-Wesley, Reading, MA (1974)

3 H. Haken, Light, Vol. 2, Laser Light Dynamics, North Holland, Amsterdam (1985)

4 H. Risken, K. Nummedal, J. Appl. Physics 39, 4662 (1968)

5 H. Risken, C. Schmid, W. Weidlich, Z. Physik, 194, 337 (1966)

5a N.S. Goel, N. Richter-Dyn, Stochastic Models in Biology, Academic Press, New York (1974)

6 H. Haken, Advanced Synergetics, Springer Series in Synergetics, Vol. 20, Springer, Berlin (1983)

7 E.N. Lorenz, J. Atmos. Sci. 20, 130 (1963)

8 H. Risken, K. Nummedal, Phys. Lett. 26A, 275 (1968)

9 R. Graham, H. Haken, Z. Physik 213, 420 (1968)

10 G.E. Forsythe, W.R. Wasow, Finite-Difference Methods for Partial Differential Equations, Wiley, New York (1967)

11 P.R. Gerber, M. Büttiker, Z. Physik B33, 219 (1979)

12 M. Mayr, H. Risken, H.D. Vollmer, Optics Comm. 36, 480 (1981)

13 A.P. Kazantsev, S.G. Rautian, G.I. Surdutovich, Sov. Phys. JETP 27, 756 (1968)

14 L.A. Lugiato, P. Mandel, S.T. Dembinski, A. Kossakowski, Phys. Rev. A18, 238 (1978)

15 H. Knapp, H. Risken, H.D. Vollmer, Appl. Phys. 15, 265 (1978)

16 J.C. Antoranz, M.G. Velarde, Optics Comm. 38, 61 (1981)

17 E. Arimondo, F.Casagrande, L.A. Lugiato, P. Glorieux, Appl. Phys. B30, 57 (1983)

18 L.W. Casperson, J. Opt. Soc. Am. B2, 993 (1985)

19 L.A. Lugiato, L.M. Narducci, E.V. Eschenazi, D.K. Bandy, N.B. Abraham, Phys. Rev. A32, 1563 (1985)

20 R. Bonifacio (ed.), Dissipative Systems in Quantum Optics - Resonance Flurorescence, Optical Bistability, Superfluorescence, Springer, Berlin (1982)

21 L.A. Lugiato in: Progress in Optics XXI, p.69, Ed. E. Wolf,

Elsevier Science Publisher, Amsterdam (1984)

22 L.A. Lugiato, R.J. Horowicz, G. Strini, L.M. Narducci, Phys. Rev. A30, 1366 (1984)

23 K. Ikeda, Optics Comm. 30, 257 (1979)

24 L.W. Hillman, J. Krasinski, R.W. Boyd, C.R. Strout Jr., Phys. Rev. Lett. 52, 1605 (1984)

25 L.M. Narducci, J.R. Tredice, L.A. Lugiato, N.B. Abraham, D.K. Bandy, Phys. Rev. A33, 1842 (1986)

INSTABILITY ANALYSIS OF CO_2 LASER WITH FEEDBACK CHANGING OF CAVITY LENGTH

Li-Xue Chen, Chun-Fei Li, Qiang-Sheng Hu and Jie-Fei Li

Department of Applied Physics, Harbin Institute of Technology,
Harbin, The People's Republic of China

In the last years, the investigations of instability in laser system have attracted a great deal of attention [1]-[4]. Many different references consider chaotic behaviours of a signal mode CO_2 laser. The CO_2 laser is a homogeneous broadening laser, the adiabatic elimination of the polarization can be satisfied. The third equation can be introduced and a chaotic system of three first-order nonlinear autonomous differential equations is formed. This paper proposes a way to introduce the third differetial equation for the laser instability and the instability analysis of the system is done for research of the chaotic behaviours.

The experimental setup of proposed chaotic model is shown in Fig. 2. This is that a part of the output of a laser is converted into the amplified electrical signal which is applied to PZT to adjust the cavity length. The changing of the cavity length L is described by the Debye relaxation equation as follows:

$$\tau(dL/dt) + L - L_0 = \alpha I \qquad (1)$$

where I is the intensity of the laser output, L is the length of the laser cavity, L_0 is the length of the cavity without feedback. τ is the Debye relaxation time of feedback system, α is the conversion coefficient including one of optico-electric and piezoelectric converse effect .

According to the definition of laser detunig and relationship between the detuning δ and the cavity length L, the detuning and the length satisfy the relation as follows,

$$\delta = (\omega_c - \omega_a) / \gamma_\perp = (2\pi mc/L - \omega_a)/\gamma_\perp \qquad (2)$$

where ω_c and ω_a are the cavity and the atomic frequencies, respectively. $\gamma_\perp$ is the transverse relaxation rate, m is the longitudinal mode number of the laser. c is the light speed.

From Eqs. (1) and (2), the differential equation of the detuning can be obtained. Then, with addition of two first-order nonlinear differential equations in CO_2 laser system after adiabatic elimination of polarization, the system is described by the following equations[4],

$$dI/dt = -2kI + 2RID/(1 + \delta^2) \qquad (3a)$$

$$dD/dt = \gamma_{\parallel}[1 - D - ID/(1 + \delta^2)] \qquad (3b)$$

$$\tau(d\delta/dt) = (\delta + \omega_a/\gamma_{\parallel})\{ 1 - (\delta + \omega_a/\gamma_{\parallel}) [1/(\delta_0 + \omega_a/\gamma_{\parallel}) + \alpha\gamma_{\parallel}I/2\pi mc] \} \qquad (3c)$$

where D is the population inversion, δ_0 is the initial detuning without feedback. R is the pump rate, k is the attenuation coeffcient of laser cavity, $\gamma_{\parallel}$ is the longitudinal relaxation rate。

A linearized stability analysis of the equations gives out the eigenvalue equation as follows

$$\lambda^3 + a_1\lambda^2 + a_2\lambda + a_3 = 0 \qquad (4a)$$

where

$$a_1 = 1/\tau + \frac{\gamma_{\parallel}(R/k)}{(1 + \delta_s^2)} \qquad (4b)$$

$$a_2 = \frac{[\tau^{-1}\gamma_{\parallel}(R/k) + 2k\gamma_{\parallel}I_s]}{(1 + \delta_s^2)} - \frac{4k\alpha\gamma_{\perp}\tau^{-1}}{(1 + \delta_s^2)}\frac{(\delta_s + \omega_s/\gamma_{\parallel})^2}{2\pi mc} I_s\,\delta_s \qquad (4c)$$

$$a_3 = \frac{2k\gamma_{\parallel}\tau^{-1}I_s}{(1 + \delta_s^2)} - \frac{4k\alpha\gamma_{\perp}\tau^{-1}\gamma_{\parallel}}{(1 + \delta_s^2)}\frac{(\delta_s + \omega_a/\gamma_{\parallel})^2}{2\pi mc} I_s\,\delta_s \qquad (4d)$$

where I_s and δ_s are the intensity and the detuning in the case of stable state, respectively. Suppose $\lambda = ib$ is a complete imaginary number and it is substituted into Eq.(4a), the instability conditions are obtained,

$$b^2 = a_2 = a_3/a_1 \qquad (5a)$$

$$a_2 > 0 \qquad (5b)$$

$$a_3 / a_1 > 0 \qquad (5c)$$

Substituting Eqs.(4a) - (4c) into Eqs.(5a) - (5c), one can obtain the instability condition as follows

$$\tau^{-1} > \gamma_{\parallel} \qquad (6a)$$

$$I_s > \frac{\tau^{-1}\gamma_{\parallel}(R/k)}{2k(1/\tau - \gamma_{\parallel})} \qquad (6b)$$

Eq.(6a) shows that the instability can appear only when the Debye relaxation rate is larger than the longitudinal relaxation rate. Considering the relationship between the intensity I_s and pump rate R in the case of stable state the instability condition Eq.(6a) can be rewritten in the following form

$$I_s > \frac{\tau^{-1}\gamma_{\parallel}}{2k(1/\tau-\gamma_{\parallel})}\{1+I_s+[\frac{1}{(1/(\delta_s+\omega_0/\gamma_{\parallel}))+(\alpha\gamma_{\perp}I_s/2\pi mc)}-\frac{\omega_0}{\gamma_{\perp}}]^2\} \tag{7}$$

Eq.(7) shows that the instability can appear only when the intensity I_s

is falled into a region . Fig. 2(b) and (c) show the variation of the instability region with Debye relaxation time τ on the I_s-R curve.

The numerical and experimental research about the chaotic model are being carried out, the results will be published in other paper.

REFERENCES

[1] N. B. Abraham, M. D. Coleman, M. Maeda, and J. Wesson, Appl. Phys. B 28, 169(1982).

[2] C. O. Weiss and H. King, Opt. Commun., 44, 59(1982)

[3] F. T. Arrechi, R. Meucci, G. P. Puccioni and J. R. Tredicce, Phys. Rev. Lett., 49, 1217(1982).

[4] Li-Xue Chen, Ai-Qun Ma and Chun-Fei Li, Optica Sinica, 7 to be published (1987).

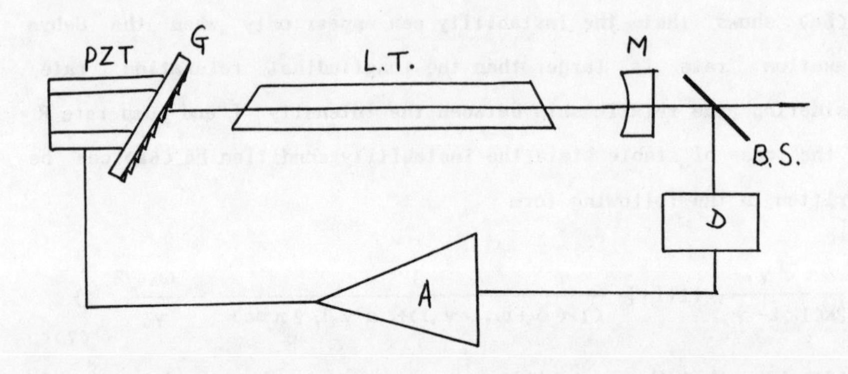

Fig. 1 Experimental setup of CO2 laser with feedback changing of the cavity length. D – HgCdTe detector; B.S. – beam splitter, L.T. – CO2 laser tube, B.V. – bias voltage, A – DC amplifier, M – mirror, PZT – converse piezoelectric crystal.

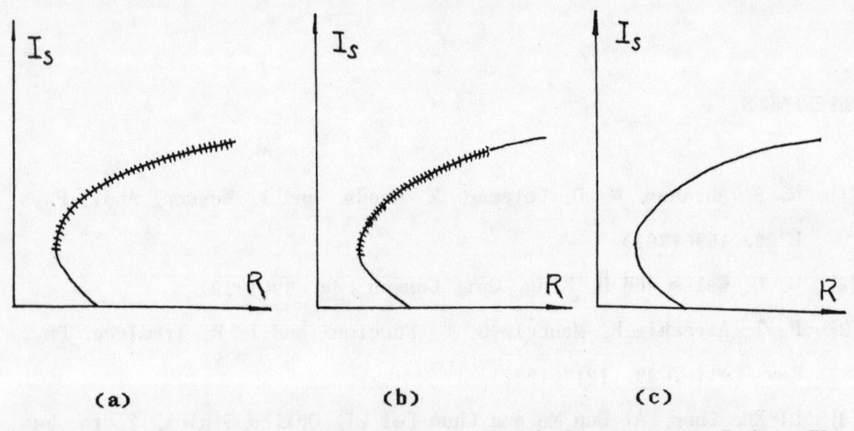

Fig. 1 Relation curve of the intensity I_s and the pump rate R for CO2 laser with feedback changing of cavity length. (a) The bistable and monostable curve; (b) larger instability region for $\tau^{-1}=2.51\times10^3$, $\alpha=8.0\times10^{-4}$; (c) instability region for $\tau^{-1}=1.95\times10^3$, $\alpha=8.0\times10^{-4}$

CHAOS CONTROL THEORY OF LASER SYSTEMS

Li Fuli and Cao Zexian

University of Science and Technology of China, Hefei, China

ABSTRACT

It has been shown that one could have controls of 'chaos to order', 'order to chaos' and 'chaos-order-chaos' series by using intensity correlation functions as feedback signals.

25/6/1987

Prepared for The International Workshop On Instability Optical Bistability and Optical Computers, Beijing, 1987

The dynamics of chaos is an active research field. The theoretical activities have envolved from the pioneering works of Lorenz Haken[1][2]. It has been shown that chaos is by no means an exceptional but an universal behaviour.[3] Many chaotic systems have also been studied experimentally. It is natural to ask whether chaos theory could find any technical application. From this point of view a novel control theory of chaotic systems might be of interest for the future chaos technology. Recently one of the authors(LFL) has presented a correlation control theory of chaotic laser systems in which the field correlation functions are used as feedback signals.

In this paper computer experiments have shown that it is also possible to have controls of 'chaos to order', 'order to chaos' and 'chaos-order-chaos' series based on intensity correlation control theory.

As it is well-known that dozens of systems can be modeled by Lorenz equations

$$\begin{aligned} \dot{x} &= -(x-y) \\ \dot{y} &= -xz-y-rx \\ \dot{z} &= xy-bz \end{aligned} \qquad (1)$$

For a single mode homogeneous laser system here $\sigma = \kappa/\gamma_\perp$, $b = \gamma_\parallel/\gamma_\perp$, $r = D_0/D_t$ and the time unit is in $\gamma_\perp$. For $r=1$, the trivial solutions $x=y=z=0$ becomes unstable, and one observes the emergence of two fixed points

$$\begin{aligned} z &= r-1 \\ x = y &= \pm\left[b(r-1)\right]^{1/2} \end{aligned} \qquad (2)$$

which turn to be stable, if $\sigma > (b+1)$ and

$$r > r_c = 1 - (\sigma - b - 1)(\sigma - 1)/[b(\sigma - b - 1)]$$

all the stationary solutions are unstable. The critical point r_c marks the onset of chaos.

We use the intensity correlation functions as feedback signal in

$$\dot{x} = -(x-y)$$
$$\dot{y} = -xz - y - r_* G_* x$$
$$\dot{z} = xy - bz$$

(3)

where

$$G = \begin{cases} 1 & t \leq T \\ f(g^{(2)}(\tau) & t > T \end{cases} \qquad (4)$$

$$g^{(2)}(\tau) = \langle E^2(t+\tau)E^2(t)\rangle / \langle E^4(t)\rangle \qquad (5)$$

here $f(g^{(2)}(\tau))$ is the function of $g^{(2)}(\tau)$. For the simplest case one could use $f(g^{(2)}(\tau)) = g^{(2)}(\tau)$. We take $\sigma=3$ and $b=1$ in the following.

Fig.1 shows the intensity correlation function $g^{(2)}(\tau)$ for r=22 It is obvious that for a small τ one could have $r\, g^{(2)}(\tau) < r_c$. For general cases one should use a function of g ( ) to ensure that $r\, f(g^{(2)}(\tau)) < r_c$.

A correlation control process is shown in Fig.2, for $r=22>r_c$. As can be seen that for $t<=T$ it has chaotic behaviour and when $t>T$ it is going to have time-independent output while $r>r_c$.

The 'chaos-order-chaos-order' series is shown in Fig.3 for r=22. In this case a periodic feedback is used in the following way

$$G=\begin{cases} 1 & (n-1)T < t \leqslant nT \\ f(g^{(2)}(\tau)) & nT < t \leqslant (n+1)T \end{cases} \qquad n=1,\ 2,\ 3---$$

One can also make that $r\ f(g^{(2)}(\tau))>r_c$ when $r<r_c$. In this case one has an 'order to chaos' control process as shown in Fig.4.

We have also calculated the control processes for inhomogeneous broadened chaotic laser systems. And the correlation control approach can also be extended further more. For example one can use other characters of chaos as feedback signals. The problems of experimental test of the correlation control theory will be discussed elsewhere.

In conclusion we should mention that the chaos control theory may one day be of importance not only for the future chaos technology but also for life science, brainscience, social science and thinking processes. For example there are evidences which show that the brain can exbit chaos. Economical systems have also chaos behaviours. The chaos control theory may play a crucial role in these complex systems.

REFERENCES

(1) E.N.Lorenz, J. Atom. Sci. 20 (1963) 130

(2) H.Haken, Phys. Letters 53A (1975) 77

(3) Hao Bailin, 'Chaos', Singapore World Scientific Publishing 1984

(4) Li Fuli, ICTP Internal Report IC/86/53, ICTP, Trieste, Italy

(5) Li Fuli, 'Chaos Laser and Chaos Control Theory' National Conference On Laser Spectroscopy(Hefei, China, 1986)

(6) Niu Pengfei and Li Fuli, 'Correlation Control Theory Of Inhomogeneous Broadened Chaotic Laser Systems',Accepted by The International Laser Conference (Xiamen, China, 1987)

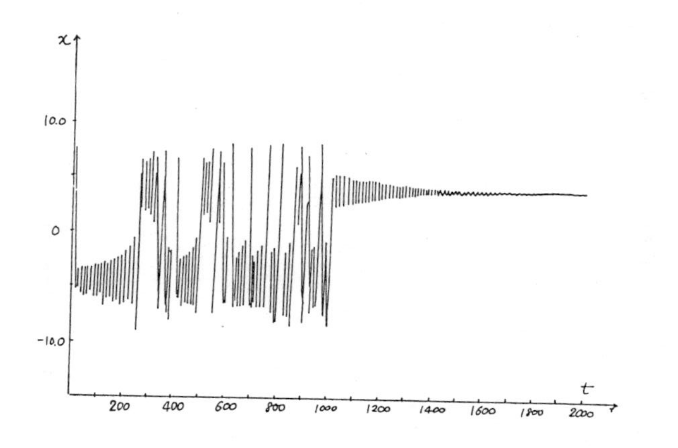

$g^{(2)}(\tau)$
2.0
1.5
1.0
0.5
100
200
300
400
500
τ

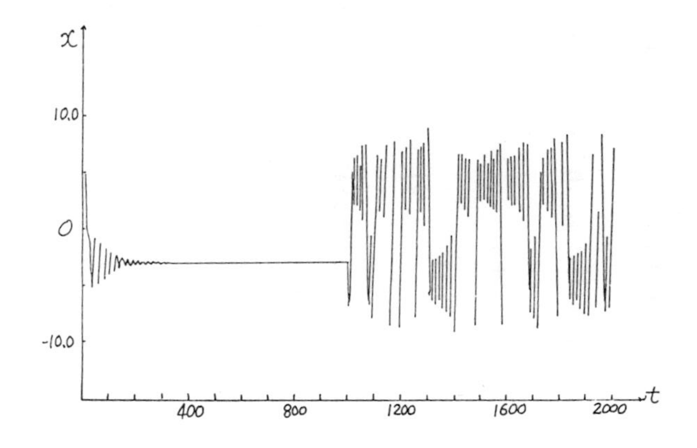

x
10.0
0
-10.0
200
400
600
800
1000
1200
1400
1600
1800
2000
t

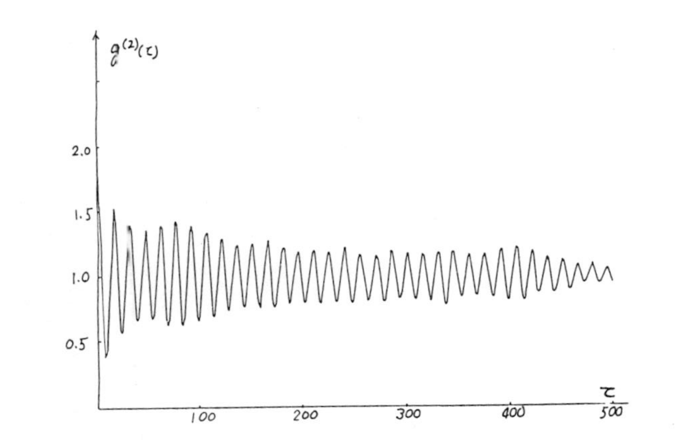

10.0
0.5
0
-5.0
800
1600
2400
3200
4000
t

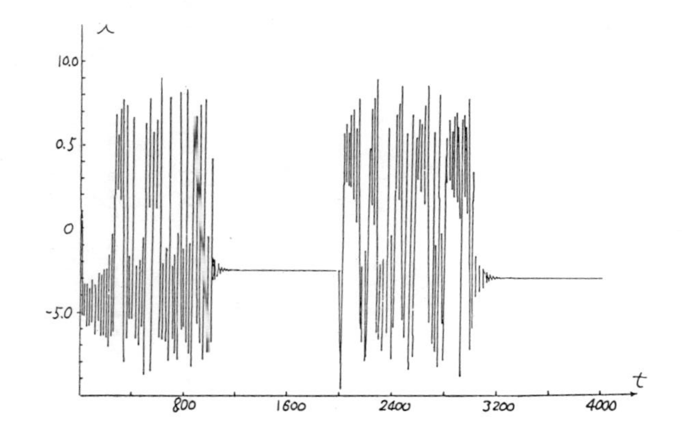

x
10.0
0
-10.0
400
800
1200
1600
2000
t

Frequency-locking structure and the Farey sequences in optical bistability with competing interactions

Fu-lai Zhang, Shi-ping Yang, Guang Xu,

Jian-hua Dai and Hong-jun Zhang

(Institute of Physics, Chinese Academy of Sciences,

P. O. Box 603, Beijing, China)

ABSTRACT

Frequency-locking structure and the Fare-fraction parameter sequences in optical bistability with competing interactions are investigated. The hole structure and the relationship between the hole and the frequency jump is predicted in the frequency locking structure. Some related experimental results obtained from a hybrid optical bistable device with two delayed-feedback loops are presented.

In recent years, optical bistability with competing interactions has attracted much attention.[1-3] Ikeda and

Mizuno[4] first studied instabilities in a nonlinear Fabry-Pebry resonator and an equivalent nonlinear ring resonator with two component cavities. A competition between the time-delayed feedback cause a "frustration" in selecting an oscillation mode. The oscillation frequency jumps discontinuously and at random as the ratio of delay times $w=|t_1-t_2|/(t_1+t_2)$, where t_1 and t_2 are two delay times, is varied.

Zhang and Dai[5] have observed experimentally the instability in a hybrid liquid crystal optical bistable device with two time delay, including the frequency-locking, jump between frequency-locking step, hysteresis and frequency-beating pattern due to competing interactions.

In this paper, the relation of oscillation frequency and instability threshold to parameter w are estabilished with the help of linear stability analysis near the instability threshold. The frequency-locking structure and the Farey-fraction parameter sequences are investigated. Some related experimental result obtained with a hybrid optical bistale device are also presented.

The dynamics of the hybrid optical device with two delayed feedback loop is described by the differential equation

$$dX(t)/dt=-X(t)+A\sin^2[X(t-t_1)+X(t-t_2)-X_b] \qquad (1)$$

where A, X(t), and X_b are proportional to the input intensity, the output intensity, and the bias voltage, respectively; t_1, and t_2 are the two delayed times of the feedback loops, and all time variables are scaled to the relaxation time of the system. The steady state of the system (denoted by X_s) obeys the follwing equation

$$X_s=A\sin^2(2X_s-X_b) \qquad (2)$$

The stability of steady state is controlled by

$$\lambda=-1+A\sin[4X_s-2X_b](\exp(-\lambda t_1)+\exp(-\lambda t_2)) \qquad (3)$$

where eigenvalue $\lambda=\alpha+\beta$. If $\alpha>0$, the oscillation mode, corresponding to the eigenvalue λ, is unstable, while the imaginary part measures the oscilllation frequency. For long delay case (i.e. $\tau=(t_1+t_2)>>1$), we find that the oscillation modes at the vicinity of instability have frequencies $\beta_q=2\pi q(1-2/\tau)/\tau$ specified by positive integers q. The steady state becomes unstable and an oscillation is estabilished as A increased to a certain threshold values. It was noted that the first unstable eigenmode of the system (denoted by q^*) must be correspond to the minimum A[4,6] (denoted by A^*). From eq.(3), we have

$$(A\sin(4X_s-2X_b))^{-1}=2(-1)^q\cos(2\pi q/\tau)\cos(w\pi q(1-2/\tau)) \qquad (4)$$

the right side of eq.(4) is similar to the net gain function of Ref.4.

We can take the maximum oscillation mode q_{max} so that $\cos(2\pi q/\tau)$ takes its first zero-point, that is, $q_{max}=[\tau/4]$, here [] denotes to take integer.

For a given τ, X_b and parameter w, q^* can be determined as follows: first, one gets all A for $q=1,2,3,...,q_{max}$, then, selectes the mode with minimum A among them. The oscillation mode and the minimum A are the q^* and the A^*, respectively.

Figures 1(a)-1(c) show the relations of $A^*, q^*, (q^*)^{-1}$ vs parameter w, respectively. From Fig.1(b)-1(c), we observed that the oscillation mode is a single-value function of w. For some values of w, small fluctuation of w forces the oscillation mode of the system to jump from one to another. We also observed that a series of locking steps appear in a small region of w. For example, for w=1/2,2/3,3/4,..., and 1/2,1/4,1/6,..., the locking steps form two families. The location of these steps are characterized by $w_s=P_s/Q_s$ and the oscillation mode of these locking steps is equle to Q_s. Whole q^*-w (or $(q^*)^{-1}$-w) curve consists of locking steps. Both the height and width of the step are related to w. The higher the oscillation mode, the narrower the width. The families of step are indicated by arrows in Fig.1(c). For a given family, the width of each step can be expressed as $w_i=c_i/q^*+D_i$, where i denotes a certain family, q^* is the oscillation mode of the step, C_i and D_i are constants. For families (I) (i=1) and (II) (i=2) in Fig.1(c), $C_1=0.09599$, $D_1=1.98\times10^{-3}$, $C_2=0.0975$, $D_2=-2.75\times10^{-3}$.

In addition to the frequency-locking steps, there are also many "holes" [6] in Fig.1(b). A hole is a very narrow region of the parameter w with one lowest oscillation mode in it. The location of the hole is characterized by the parameter w at the bottom. These values of w and their rational expression P_h/Q_h are listed in table (I). Their position are indicated by b_1, ..., h_1 and b_1, c_2, d_2, ..., h_2 in Fig.1(b). From table (I), we see that the parameter sequences corresponding to the positions of b_1, c_1, ..., h_1 and b_1, c_2, ..., h_2 are 1/3, 1/5, 1/7,..., 1/15 and 1/3, 3/5, 5/7, ..., 13/15, respectively. We find that the sum of P_h and Q_h is even. Any two consecutive fractions p/q and r/s of these two sequences do not satisfy the unimodularity condition $|ps-qr|=1$.[7] Usually, the quantity $|ps-qr|$ is called the modularity of p/q and r/s. In this sense, modularity of the former sequence is equal to 2 and the latter is 4. Thus, unimodularity gaps are to be introduced. Similar unimodularity gaps were also observed in many different systems[8-10]. The location of these two sequences on the Farey tree are shown in Fig.2. The former sequence is characterized by ◯, and the latter by ◌.

We see from Fig.1(b) that a series of locking steps go up or go down in the vicinty of a hole (charecterized by $w_h=P_h/Q_h$). These steps can be characterized by fractions $w_s=P_s/Q_s$. The oscillation mode of a step is equal to Q_s (i.e. $q^*=Q_s$). We find that the sum of the P_s and Q_s is odd. We also find that the interval of jump of the oscillation modes in the neighboring steps is equal to Q_h (i.e. $\Delta q^*=Q_h$). When $q^*>Q_s=[\tau/4Q_h]$ ([ ]dnotes to take integer), the oscillation modes fall into the hole abruptly. We note from Fig.1(a) that the instability thresholds of the oscillation mode in the holes are relatively higher, thus some peaks appear.

Our experiments were carried out with a hybrid optical bistable system designed with two delayed-feedback loops. The experimental setup is shown in Fig.3. The output waveforms and their power spectra were measured by a dual-channel fast-Fourior transform(FFT) analyzer. In order to get some experimental results for large τ and adapt the output oscillation frequency to our FFT, a $LiNbO_3$ crystal modulator was used. Its half-wave voltage is about 566 V. Bias voltage V_b is about 300 V. So, $X_b \approx -1$. The relaxation time of the the system is approximately 5.5 msec. By selecting different buffer lengths of the microcomputer, we have adjusted the two delayed times, i.e. the parameter w.

Fig.4 shows the typical experimental waveform and its power spectrum for $\tau \approx 130$, w=0.31934, where the fundmental frequency is 11.125 Hz. Both the theoretical and the experimental curves of oscillation frequency vs w, with $0<w<1$ and $\tau \approx 130$, are given in Fig.5. The experimental results are in agreement with our theory.

In order to further verify the existence of the holes in the frequency-locking structure, as shown in Fig.1(b), we observed the relation between the oscillation frequency to the parameter w in our system for τ =1000. Fig.6 shows the result for $0.28<w<0.38$ and $X_b \approx -1$. We note that our experimental result is quite in agreement with the theory.

In conclusion, we have obtained frequency-locking structure of a hybrid optical bistable system with two time delayed feedbacks. We found the relationship between the structure and the Farey sequences. Althrough the Farey sequences appear in the structure, the physical significance is a intersting problem to be explored.

References

[1] S. L. McCall, Appl. Phys. Lett. 32, 284(1978)

[2] M. Okada and K. Takizawa, IEEE J. Quantum Electron. QE-17, 517(1981)

[3] M. M. Cheung, S. D. Durbin and Y. R. Shen, Opt. Lett. 8,39(1981)

[4] K. Ikeda and M. Mizuno, Phys. Rev. Lett. 53, 1340(1984)

[5] H. J. Zhang and J. H. Dai, Opt. Lett. 11, 245(1986)

[6] K. Ikeda, Private communication

[7] G. H. Hardy and E. M. Wright, An Introduction to the Theory of Numbers, 4th ed.(Claredon Press, Oxford, 1954)

[8] T. Allen, Physica (Amsterdam) **6D**, 305(1983)

[9] J. Maselko and H. L. Swinney, Phys. Scr. **T9**, 35(1985)

[10] D. D. Coon, S. N. Ma, and A. G. U. Perera, Phys. Rev. Lett. **58**, 1139(1987)

Figure Captions

Fig.1. the theoretical results for τ =1000 and X_b=-1. (a) the a*-w relation;(b) The q*-w relation;(c) the$(q^*)^{-1}$-w relation

Fig.2. The Farey tree. ◯ and ◌ correspond to the parameter values indicate by $b_1,\ldots, h_1$ and b_1, c_2, ...,h_2,respectively.

Fig.3. Schematic diagram of the hybrid optically bistable device with two delays in the feedback loops. He-Ne, laser; A, attenuater; P_1 and P_2, polarizers; NM, nonlinear medium--$LiNbO_3$ crystal; D, photodetector; AMP, amplifier; FFT, fast-Fourier-transform analyzer; A/D, analog-to-digital converter; D/A, digital-to-analog converter; BUF, microcomputer buffer; HVA, high voltage amplifier; V_b, bias voltage.

Fig.4. The typical experimental waveform and its power spectrum for $\tau \approx 130$, $w=0.31934$, and $X_b \approx -1$. The fundmental frequency is 11.125 Hz.

Fig.5. The first oscillation mode q^*(or frequency f) as a function of w for $\tau = 130$ and $X_b = -1$.

(a) theoretical result (b) experimental result

Fig.6. The experimental result of the hole structure near by $w=1/3$ for $X_b \approx -1$, $\tau = 1000$.

Table (I). The parameter values of w and their rational expression for b_1, c_1, ..., h_1 and b_1, c_2, d_2,..., h_2 of Fig.1(b).

	w values	rational expression		w values	rational expression
h_1	0.067	1/15	h_2	0.868	13/15
g_1	0.077	1/13	g_2	0.848	11/13
f_1	0.091	1/11	f_2	0.820	9/11
e_1	0.111	1/9	e_2	0.779	7/9
d_1	0.143	1/7	d_2	0.716	5/7
c_1	0.200	1/5	c_2	0.601	3/5
b_1	0.334	1/3	b_1	0.334	1/3

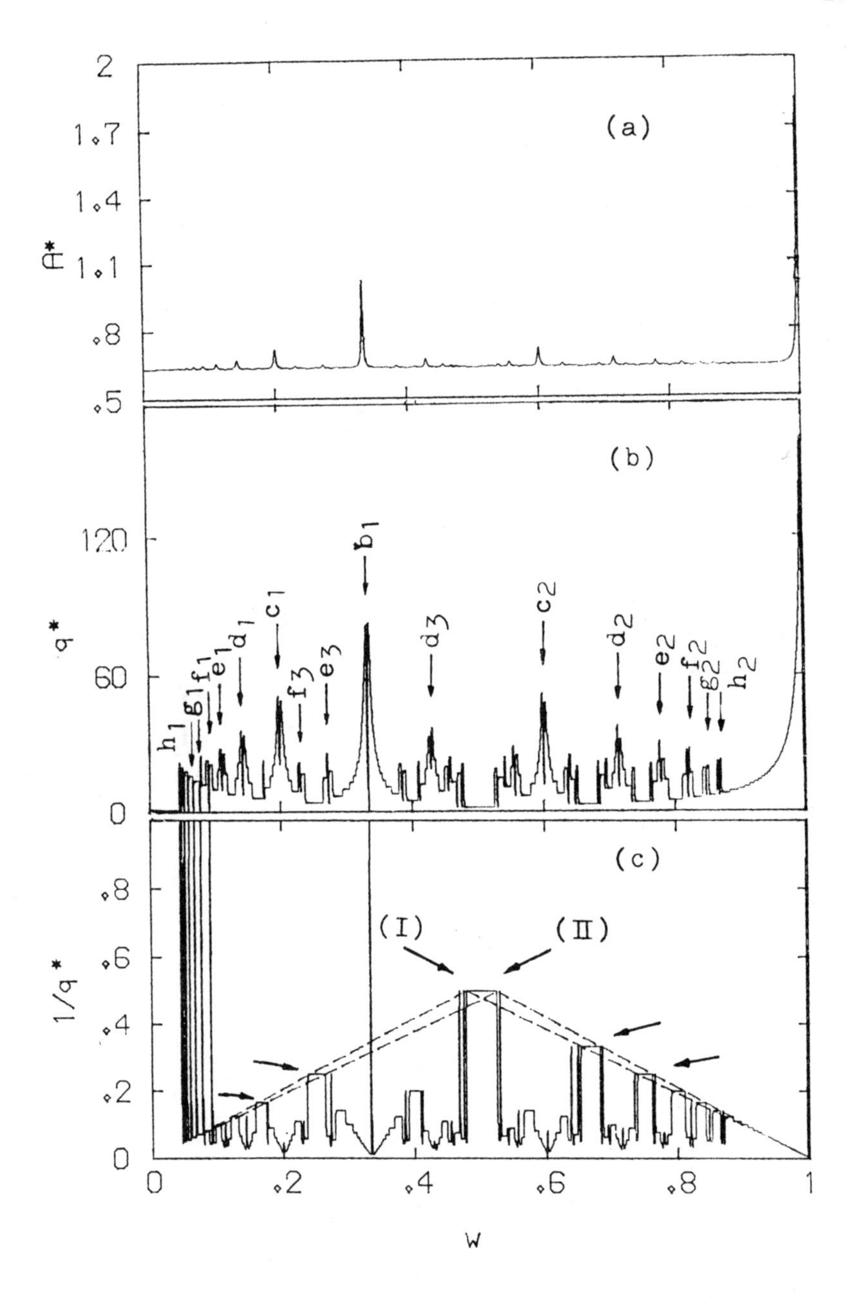
(a)
A*
2
1.7
1.4
1.1
.8
.5
(b)
q*
120
60
0
b1
c1
c2
d1
d2
d3
e1
e2
e3
f1
f2
f3
g1
g2
h1
h2
(c)
1/q*
.8
.6
.4
.2
0
(I)
(II)
0
.2
.4
.6
.8
1
W

Fig.1

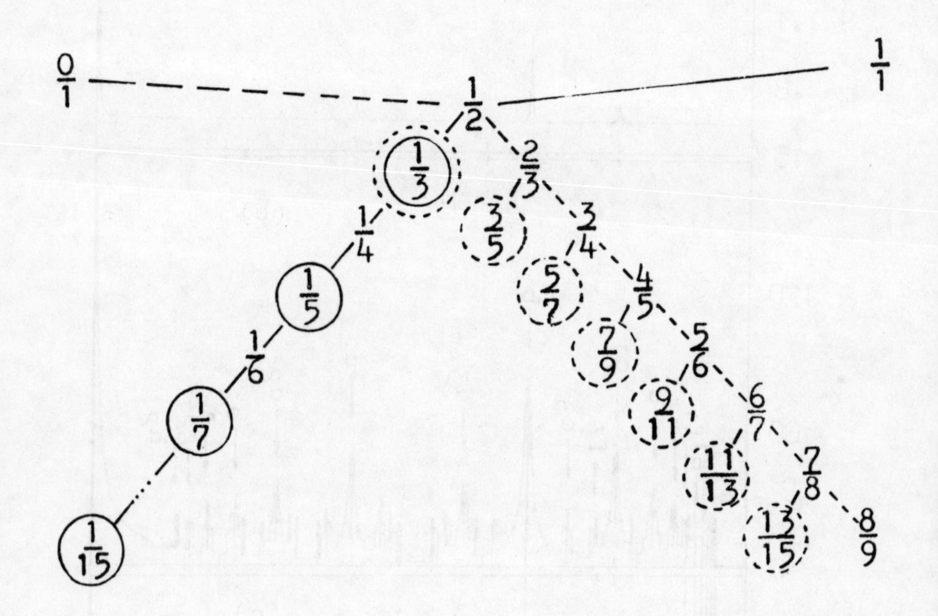
0/1
1/1
1/2
1/3
2/3
1/4
3/5
3/4
1/5
2/7
4/5
1/6
7/9
5/6
1/7
9/11
6/7
11/13
7/8
1/15
13/15
8/9

Fig.2

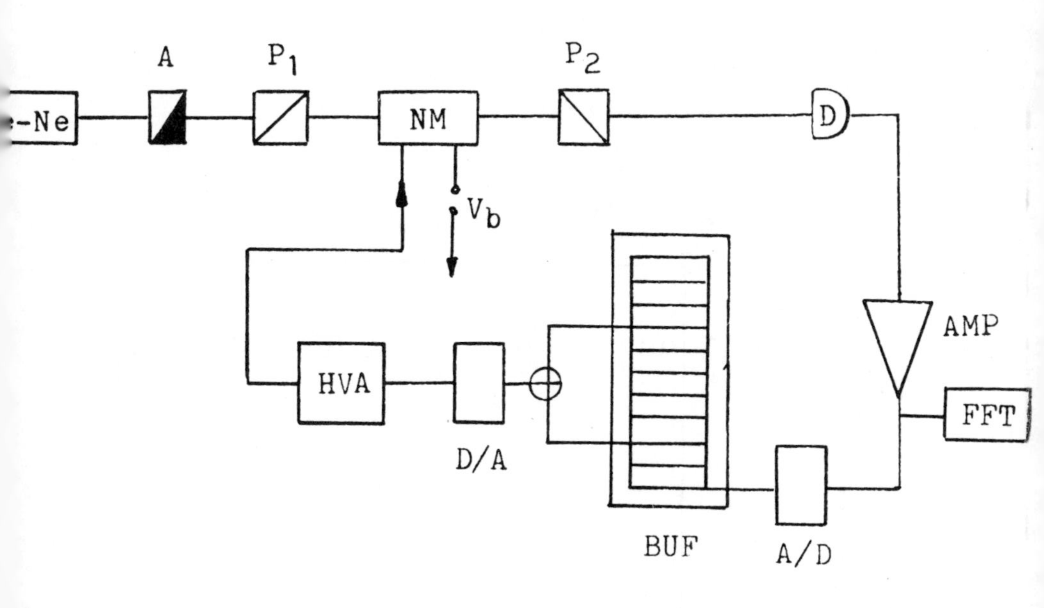

Fig.3

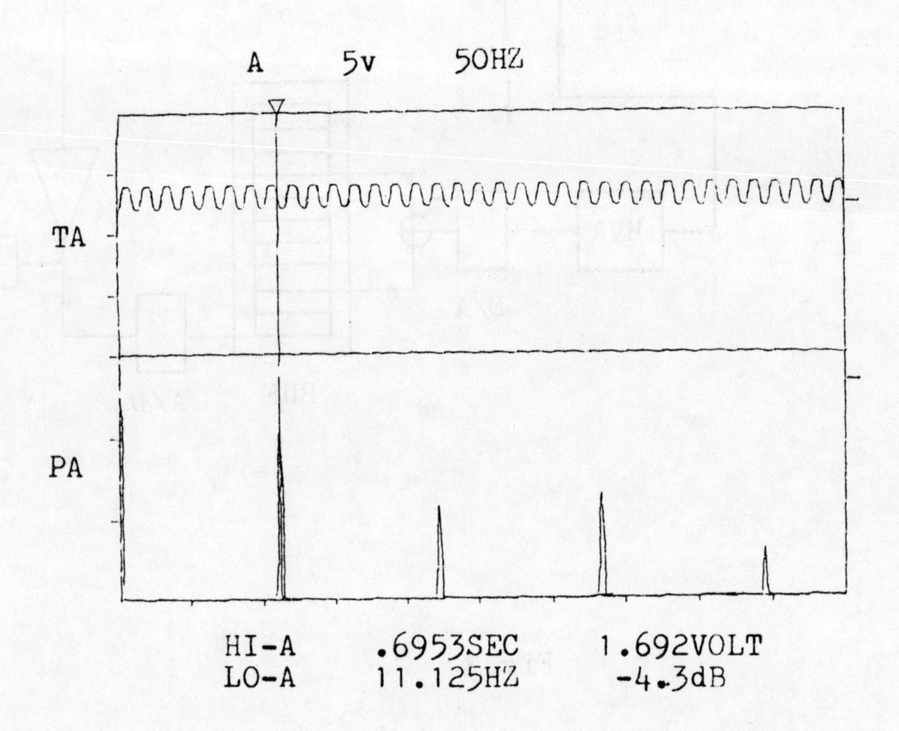

Fig.4

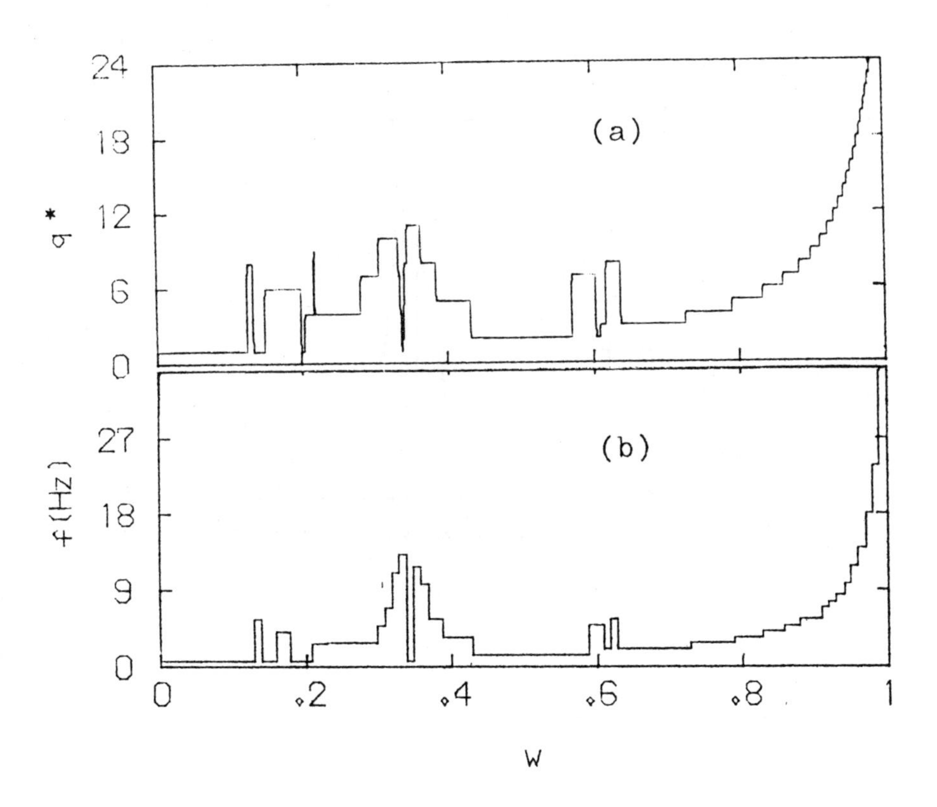

Fig.5

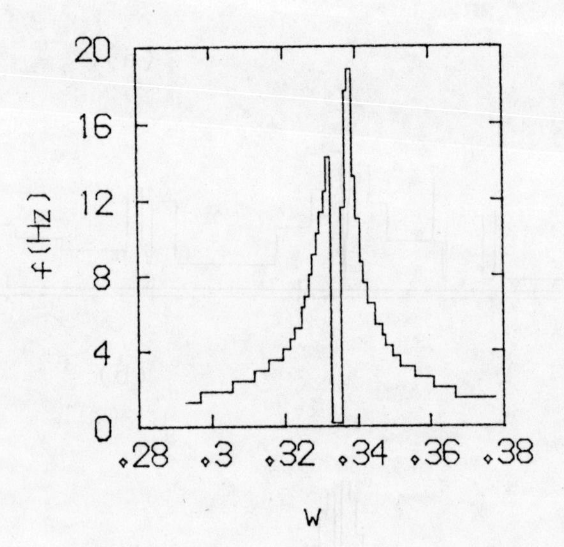

Fig.6

RESEARCH PROGRESS IN OPTO-ELECTRONICS AND OPTICAL COMPUTING AT INFORMATION ELECTRONICS DEPARTMENT OF TSINGHUA UNIVERSITY

ZHANG Keqian, GUO Yili and ZHOU Bingkun

Department of Information Electronics
Tsinghua University, Beijing, China.
Telephone: 282451 Ext.2995 Telex: 22617 QHTSC CN

ABSTRACT

Opto-Electronics is one of the most emphasized fields in this department. An Integrated Opto-Electronics (IOE) Laboratory was established closely related to the department. In recent years, some research works have been carried out in the field of optical fiber communication. An External Cavity Diode Laser with less than100 KHz linewidth and long-term frequency stability was developed for coherent optical communication. Achievements are obtained in the researches of Ti:$LiNbO_3$ Waveguide Modulator and Switching Array, Semiconductor Optical Bistable Memory Switch for high-speed, real-time optical switching (data exchange) system, Optical Fiber Components and Optical Fiber Sensor System. Some new topics such as Stationary Magnetic Wave devices and Integrated Optical Frequency Spectrum Analyzer are just begining. Since 1984 we have been engaged in research of optical bistability using opto-electronic devices and in 1986 we have realized several types of optical logic and flip-flop devices. From 1987 we have began to do some fundamental researches on Optical Parallel Processing supported by the NSF of China.

The IOE Lab's facilities include three furnances of LPE, Ion Deposition Instrument, electronic and optical measurement instruments and we will get a MBE In the near future. The IOE Lab. has many joint research projects with other Laboratories in the department such as communication,image processing and microwave Laboratory

The rasearch activities on optical bistability and optical computing in the IOE Lab. are as follows:

1. **Optical Bistable Devices and Optical Instability.**

A compact, low input power optical bistable component,consisting of a semiconductor laser diode, an optical fiber directional coupler and a photodetector, was demonstrated in 1985. A steady-state characteristics analysis for this device was published in Chinese Physics: Lasers (a publication of OSA) in 1986 and 1987. In 1987, a long fiber is combined with this device to form a oscillator. When an external modulation injects into the laser, we have successfully observed injection locking and even instability in such a system.

A novel function of converting optical pulse amplitude modulation (PAM) to pulse duration modulation (PDM) based on the optical critical slowing down effect was demonstrated and published in Optics Communications in March, 1987. It may be a potential application of optical bistability.

2. Optical Logic, Optical Flip-Flop and their potential applications.

Optical gates such as **OR, NOR, AND, NAND, XOR** and **XNOR** are all simply realized in our Lab. We got them both by heterojunction phototransistor-laser combined device and by photodiode-laser combined device. The using of negative resistance effect in HPT is really a new way to realize optical bistability and logic gates. These works were published in Electronics Letters in1986 and in IEEE J. of Quantum Electronics in 1987.

Some more complicated optical flip-flops such as R-S and J-K flip-flops are the basic and important digital computing elements in optical computer and digital processors. In 1987, using two LEDs and several PINs, some new types of clocked optical R-S and J-K flip-flops based on the optical feedback and optoelectronic coupling are demonstrated in our Lab. These works will be presented at this meeting and the annual conference of OSA in 1987. These flip-flops can be easily integrated and used as an optical memory element in optical fiber communication networks and optical digital computing systems.

Ti:$LiNbO_3$ integrated waveguide modulation devices and switching devices are developed in our Lab. and some good results have been achieved in recent years.

3. Two Dimensional Parallel Optical Computing.

Parallel computing ability is one of the most important and fascinate advantages of optical computing compared with electronic computers. However, it seems that there is not a completed system accepted by most people up to now. Of course, some systems, just like S.D.Smith's system in Heriot-Watt,H.M.Gibbs's system in Arizona and Caulfield's system in Alabama, are still interesting all of us.

Very recently , we have just began a project of optical parallel processing system based on the classical finite state machine. The parallel logic gates using spatial light modulator and spatial encoding technique are investigated.

Optical Realization of 32 - Sequence Walsh-Hadamard Transformation

Y.S. Chen, S.H.Zheng, D.H. Li and G.Z. Yang
Institute of Physics, Academia Sinica
Beijing, China

Abstract

A coherent optical experimental system for realizing a 32 - sequence Walsh-Hadamard transform(WHT) in parallel operation is established. In this system, a single holographic mask and two Fourier lenses are adopted, and the amplitude-phase distribution of holographic mak is simply derived from matrix multiplications, instead of solving a large set of equations. The experimental results show that the experimental transforming spectra are in agreement with theoretical predictions.

Introduction

It is well known to perform a linear space - invariant operations (e.g. convolutions, correlations) using the Fourier transforming properties[1] of converging lenses. However, the real situations most commonly encountered is that requiring to treat with the space - variant problems. In recent years, more attention[2,3] has been devoted to performing this kind of operations by means of optical method. Comparing with electrical computer, use of optical method is able to process informations or data in parallel operations.

One of solutions for performing space - variant operations is to use optical general transformation[4,5], which stated that for an arbitary linear unitary transform there is a coherent system composed of holographic masks(HMs) to realize it, and the required masks can be designed by a iterative method. In experimental research several optical transforms in low - sequence, such as 8 - S Walsh transform[6,7] and 16 - S Mellin transform[8] in 1 - D space, have been achieved by establishing a experimental system consisted of HMs.

Doing research on optical transformation in high - sequence (over 16) is more difficult than in low-sequence one, because in high - S the increase of number of sampling points leads to enlarging apertures of the system, thus violates the paraxial approximation. Furthermore, the dimension of a set of equations, by which the amplitude-phase distribution(APD) of HM is determined, are rapidly increasing. For instance the set of equations is of 1024 equations for a 32 - S one, which is quite tedious to be solved even with a huge computer.

In this paper, a coherent system composed of a single HM and two Fourier lenses is presented to solve those problems. With this system, it is unnecessary to consider paraxial approximation and the APD of HM is directly derived from matrix multiplications, instead of solving a set of equations. A such experimental system is established in laboratory for realizing the Walsh-Hadamard transform in 1 - D space. The experimental results are in agreement with theoretical predictions.

Design of a transforming system

Fig. 1 shows the schematic diagram of a transforming system composed of a HM and two Fourier lenses(L), where ℓ is the focal length of the lens. The propagating process of lightwave from input plane(x_1) to output plane(x_2), suppose that the process completed a given transform WHT, can be represented as :

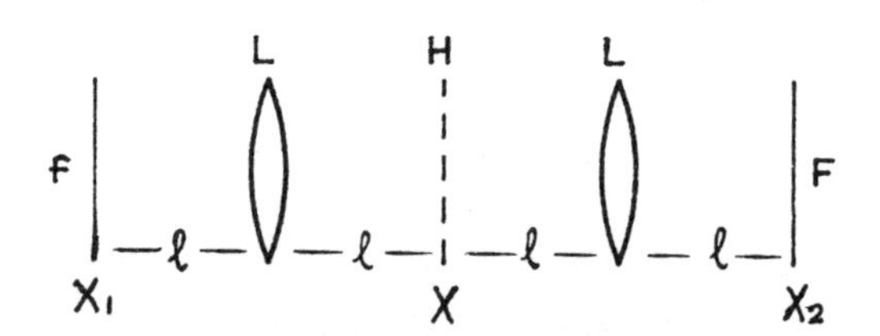

Fig. 1 Schematic diagram of a transforming system
x_1 - input plane, x - mask plane, x_2 - output plane, H - HM,
L - Fourier lens, ℓ - focal length of the lens, f - input function,
F - output spectrum.

$$F(x_2) = \int T(x_2,x_1)f(x_1)dx_1 \qquad (1)$$
$$T(x_2,x_1) = G(x_2,x).H(x).G(x,x_1) \qquad (2)$$
$$G(x_2,x) = \exp(-i2\pi x_2 x/(\ell\lambda)) \qquad (3)$$
$$G(x,x_1) = \exp(-i2\pi x x_1/(\ell\lambda)) \qquad (4)$$
$$H(x) = A(x)\exp(i\phi(x)) \qquad (5)$$

Where f and F are input function and its transforming spectrum, and $A(x)$ and $\phi(x)$ are amplitude and phase distribution respectively. T is the matrix required to perform the transformation. For a discrete sample, above expressions then become

$$F(k) = \sum_j T(k,j)f(j) \qquad (6)$$
$$T(k,j) = \sum_n G(k,n)H(n)G(n,j) \qquad (7)$$
$$G(k,n) = \exp(-i2\pi kn/N_2)\mathrm{Sinc}(rk/N_2) \qquad (8)$$
$$G(n,j) = \exp(-i2\pi nj/N)\mathrm{Sinc}(r_1 n/N) \qquad (9)$$
$$H(n) = A(n)\exp(i\phi(n)) \qquad (10)$$
$$k=1,2\ldots N_2,\ n=1,2\ldots N,\ j=1,2\ldots N_1 \qquad (11)$$

To get above reduced expressions a group of conditional relations is adopted as

$$\Delta x_1 \Delta x/(\ell\lambda) = 1/N$$
$$\Delta x \Delta x_2/(\ell\lambda) = 1/N_2 \qquad (12)$$
$$N = N_1 . N_2$$

Where Δx_1, Δx, Δx_2, and N_1, N, N_2 are sampling periods and sampling numbers in input, mask and output plane respectively. The Sinc-function as a constituent of expressions (8) and (9), is caused by the diffracting effect of size of sampling units[9]. The r_1 and r are ratio of sampling unit to sampling period in input and mask plane respectively.

It is not difficult to find that the Sinc-function factor in expression (8) can be absorbed into output plane and the other one in expression (9) can be absorbed into mask. In this way, the residues of propagating matrices $G(k,n)$ and $G(n,j)$ are of Hermitian properties, which makes $H(n)$, the APD of HM required to achieve a given transform, derived from multiplications of the relevant matrices. Hence, we have

$$H(n) = H'(n)/\mathrm{Sinc}(r_1 n/N) \qquad (13)$$
$$H'(n) = \sum_k^{N_2}\sum_j^{N_1} G^*(n,k)/\mathrm{Sinc}(rk/N_2).T(k,j).G^*(j,n)/\mathrm{Sinc}(r_1 n/N) \qquad (14)$$

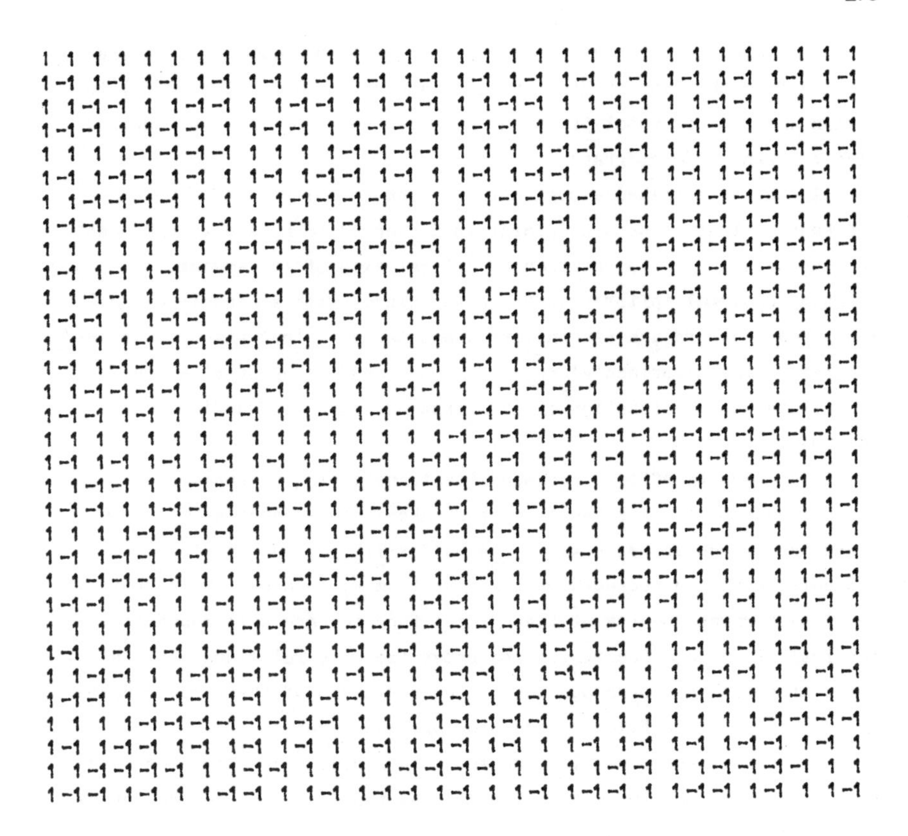

Fig. 2 The Walsh-Hadamard matrix in 32 - Sequence

Where symbol * indicates the conjugating operation. For a Walsh-Hadamard matrix T(k,j) in 32 - sequence that shown in Fig.2, we have $N_1=N_2=32$, and N=1024. By solving the equations (8),(9),(13) and (14) the H(n) is obtained. However, for the sake of brevity, the data of H(n) are not listed here.

Experiment

From the conditional relations (12) the geometric parameters of the experimental system are determined in a reasonable range. In this experiment they are taken as follows: ℓ=415mm,Δx_2=0.47mm,$\Delta x=\Delta x_1=1.47\times10^{-2}$mm, $r_1=1$, r=0.5,λ=6328A°.

The HM,which is a critical component in transforming system, was made up from the data of H(n) by using a microcomputer (IBM PC-XT) to design and plot a prototype of the HM, then reducing it to a required size by means of photograph.

In the established experimental system constituted of a HM and two Fourier lenses, each component is distanced to ℓ, and the HM must be exactly adjusted to position of optical axis of the system. A scanning photomultiplier covered on the window with a pinhole($\phi \sim 10\mu$m) mask, and a camera are alternatively mounted at output plane to detect and record the experimental results. The transforming spectra are distributed at the location of the first order of diffraction due to the HM.

In order to effectly examine the transforming performance of the system, some particular functions are taken into account as the input functions, which are 32 rows of the matrix T(k,j), but in it all -1 elements have been replaced by zero, as to make them easily in optics. The transforming spectra of these input function are well known from theoretical calculation, which indicates the spectral lines are distributed at positions corresponding to the positions of their row's number and a constant line fixed at position of number one.

The experimental results are shown in Fig.3, that shows the spectral lines are just located at the desired positions as the theoretical predictions.

Summary

A coherent system composed of a holographic mask and two Fourier lenses are proposed to perform an arbitary linear transformation in high - sequence. The 32 - sequence Walsh-Hadamard transform is realized in 1 - D space in an experimental system established in such way. The experimental results are in agreement with the theoretical predictions. In principle, the proposed system is applicable to perform a transform in 2 - D space also, where only one thing needed to do is substituting a 2 - D HM for a 1 - D one.

References

1. J.W.Goodman, "Introduction to Fourier Optics",(McGraw-Hill,New York, 1968).

2. J.F.Walkup, "Space-Variant Optical Coherent Processing", Opt. Engr., 19,339(1980).

3. W.T.Rhodes, "Space-Variant Optical System and Processing", in Applications of the Optical Fourier Transform, H. Stark, ed. (Academic Press, New York, 1981).

4. Y.P.Huo, G.Z.Yang and B.Y.Gu, "Unitary Transformation and General Linear Transformation by an Optical Method (1), the Analysis of Possibility", Acta Physica Sinica, 24,438(1975). "Unitary Transformation and General Linear Transformation by an Optical Method (2), the Iterative Method of Solution", Acta Physica Sinica, 25,31(1976).

5. G.Z.Yang, "Theory of Optical Transformation by a Single Holographic Lens", Acta Physica Sinica, 30, 1340(1981).

6. Y.S.Chen, et al., "The Optical Realization of Walsh Transformation With a Coherent System", Acta Physica Sinica, 29, 1307(1980).

7. Y.T.Wang, et al., "Optical Realisation of Walsh Transformation by a single Holographic Element", Acta Physica Sinica, 33,1599(1984).

8. S.H.Zheng, et al., "A New Method for Performing Optical Mellin Transform", Acta Physica Sinica, 35,529(1986).

9. Y.S.Chen, et al., "A Discrete Sampling Method in Optical Transform", Acta Physica Sinica, 35, 1390(1986).

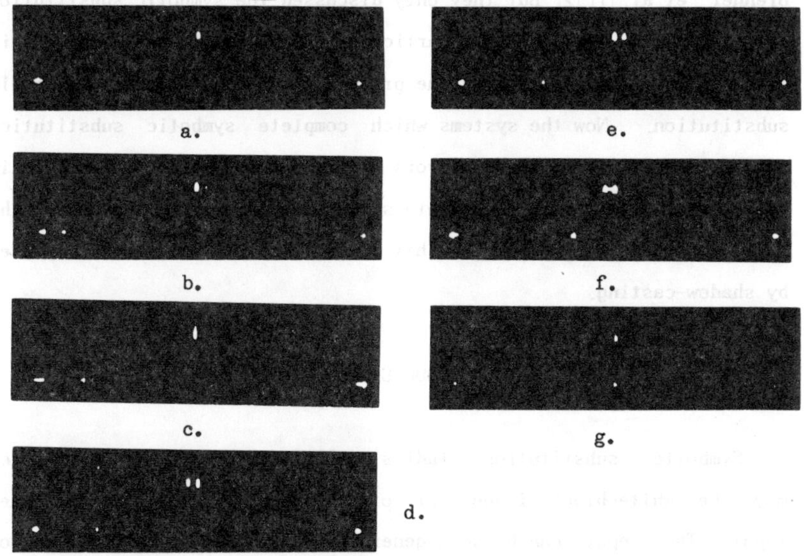

Fig.3 Experimental transforming spectra of several input functions. The spectral lines for each one are located at positions correspoding to the row's number. Photographs a, b, c, d, e, f, and g correspond to 1st, 3th, 5th, 7th, 9th, 13th, and 17th row function of the matrix.

THE GENERIC PROCEDURE AND A NEW METHOD OF IMPLEMENTING SYMBOLIC SUBSTITUTION

Wei Xue, Li-Xue Chen, Chun-Fei Li and Jing Hong

(Department of Applied Physics, Harbin Institute of Technology, the People's Republic of China)

1. INTRODUCTION

A new implementation of digital optical computing called symbolic substitution has been introduced by Huang and developed further by Brenner et al. [1] [2] But they only discussed the symbolic substitution of particular patterns in a particular input. The formulas given in these papers can not provide the procedure and conditions of symbolic substitution. Now the systems which complete symbolic substitution consisting of many prisms, mirrors and lenses are more complex, so it is necessary to seek more simple systems. We will try to discuss the generic procedure and propose that symbolic substitution be completed by shadow-casting.

2. THE GENERIC PROCEDURE OF SYMBOLIC SUBSTITUTION

Symbolic substitution studies 2-D (two-dimensional)data which may be white-black images or other abstract symbols we call them input. The input can be seen generally as a spatial permutation of patterns which have different structure . This can be expressed in terms of set theory by

$$I = \{P_i | i \in Z\}$$
$$Z = \{ i | i \text{ is natural number}\} \qquad (1)$$

Where I is the input , and P_i is the ith pattern in the input. The patterns are equal in spatial scale. The purpose of symbolic substitution is recognizing a type of patterns from the input , and substituting the patterns by others. So the object of interest in symbolic substitution is the pattern. The pattern is determined by the combination of cells which compose the pattern . Each cell only have two states, light and shade (white and black, or transparent and opaque). A pattern consisting of n cells has 2^n types of forms.

Symbolic substitution can be decomposed into recognition and subtitution phase.

(1) Recognition

Suppose an input I defined by Eq.(1)is a binary (White-black) 2-D image. We let the white be logic one. $P_s(\mathbf{R}) \in$ I is a pattern to be recognized in the input. $\mathbf{R}$ is a 2-D vector which denotes the position of a pattern. P_s consists of n cells , and the position of the ith cell whose logic value being zero is denoted by the position vector $\mathbf{S}_i$, the origin of which must be on P_s. The recognition output can be expressed as

$$I' = \{I_j' | j \in Z\}$$
$$Z = \{ j | j \text{ is natural number} \} \qquad (2)$$

The element I_j' of set I' is defined by

$$I_j' = \sum_{\alpha=0}^{N} P_i(\mathbf{R}) \overline{\underset{k=1}{\Omega}^{n-\alpha-1}} P_i^*(\mathbf{R} + \mathbf{S}_i) \qquad (3)$$

$P_j(\mathbf{R})$ is a pattern in the input, and $P_j^k(\mathbf{R}+\mathbf{S}_i^k)$ is the kth translational transform of $P_j(\mathbf{R})$, the direction and quantity of the translation are determined by $\mathbf{S}_i^k$. Ω denotes logical OR operation and $\overline{\underset{k=1}{\overset{n-\alpha-1}{\Omega}}}$ denotes the negative operation after $n-\alpha-1$ logical OR operation, $\underset{\alpha=0}{\overset{N}{\bigtriangleup}}$ denotes $N+1$ logical XOR operation. The number N in Eq.(3) is defined by

$$N = \sum_{i=1}^{n} f_i^s \qquad f_i^s \in P_s \tag{4}$$

f_i^s is the value of the ith cell in P_s. The condition of recognition is

$$I_j' = \begin{cases} 1 & \text{if } P_j = P_s \\ 0 & \text{else} \end{cases} \tag{5}$$

Here P_j and P_s denote the patterns which we only consider their structure in spite of their position, they are different from $P_j(\mathbf{R})$ and $P_s(\mathbf{R})$.

(2) Substitution

Let a substitution pattern be denoted by P_t which consists of n cells, and the value of a cell be denoted by f_i^t, the position of the ith cell the value of which is logic one is denoted by $\mathbf{T}_i^k$, the origin of T_i^k must be on P_t. The substitution output can be expressed as

$$I'' = \{I_j'' \mid j \in Z\}$$
$$Z = \{\, j \mid j \text{ is natural number} \,\} \tag{6}$$

The element I_j'' of set I'' is defined by

$$I_j'' = P_j(\mathbf{R}) \underset{k=1}{\overset{M-1}{\Omega}} P_j^k(\mathbf{R}+\mathbf{T}_i^k) \tag{7}$$

$P_j(\mathbf{R})$ is a pattern in I', and $P_j^k(\mathbf{R} + \mathbf{T}_k^j)$ is the kth translational transform of $P_j(\mathbf{R})$. The number M in Eq.(7) is defined by

$$M = \sum_{i=1}^{n} f_i^j \qquad f_i^j \in P_j \tag{8}$$

If the recognition or substitution pattern consists of four cells, it has 16 types of forms. According to the theory above, the optical processor for performing the symbolic substitution of the pattern must be able to produce four, three and two translational overlapping images of an input and perform logical NOR and logical XOR operation. If the number of the cells in a pattern increases, the number of the multi-images required in recognition and substitution also greatly increases , so the processing system for symbolic substitution will be complex. It is necessary to seek new methods and more simple systems to perform the symbolic substitution.

3. SYMBOLIC SUBSTITUTION USING SHADOW—CASTING

Shadow-casting which was first proposed by Tanida and Ichioka[3] is a method of implementing optical parallel pattern logic and optical computing. According to the conclusion above, the symbolic substitution of pattern (see Fig.1.) consisting of four cells requires four translational overlapping images of input image. These can be done by the lensless shadow-casting system. Because the shadow-casting system can not be cascaded, it is necessary to add a sequential logic gates array and a record set. Sequential logic gate has double function, memory and logical operation. A record set is used in order to record the results after recognition. The recognition and substitution phase are illustrated in Fig.2. In substitution phase, the input, middle screen, mask and logic array are removed. The result recorded after recognition is taken as an input, making use of the shadow-casting system once more, we obtain the substitution result on the output screen.

A LED (light-emitting diode)array consisting four LED's is used as a light source. Divergent light beams radiating from the LED's illuminate the input plane and project multiple shadowgrams (multi-images)of input onto the middle screen. Choosing the spacing between the LED's and distances from the source plane to the input plane and from input plane to the middle screen properly, we can make shadowgrams of input projected by the individual LED's be superposed on the middle screen, shifting one another by an amount of one cell size along the vertical and horizontal directions, the numbers and location of the multiple images is controlled by the combinations of the spatial positions of the LED's switched to the on state, so the on-off state of LED's depends on the structure of recognition and substitution patterns. Each cell in patterns only have two state, white (transparent)or black (opaque).We take the white cell as logic one and the black cell as logic zero. We define that the superposition operation of cells is logical AND in recognition and logical OR in substitution.

Suppose we had an input image which consists of two by two patterns, as shown in Fig.3., and suppose we want to recognize the pattern which is located on the lower left corner of input, i.e., the pattern in Fig.1(a), and substitute the pattern in Fig.1(b) for the pattern in Fig.1(a). The procedure of operation is as follows,

(1) Recognition

Firstly, making the four LED's (α, β, γ, δ) be in the on state, we get the four projections of input which overlap on the middle screen. Behind the mask, there are the outputs (logic one)which enter the corresponding cell of sequential logic array only at the positions corresponding to the pattern of Fig.1(c) consisting of four logic one.

Secendly, according to the position of logic one in the pattern of Fig.1(a), making the three LED's (α, γ, δ) be in the on state, we get the three productions of input. Behind the mask, there are the outputs (logic one) which enter the corresponding cell of logic array only at the positions corresponding to the patterns of Fig.1(a)and Fig.1(c) respectively.

The two outputs of the two steps above enter the logic array sequetially, and logical XOR operation can be performed. There is only one output which corresponding to the pattern of Fig.1(a)behind the logic array. the output can be recorded by record set, the result is in Fig.4.

(2) Substitution

According to the position of logic one in substitution pattern (see Fig.1(b)), making the two LED's(β, γ)be in the on state, we obtain the substitution result (see Fig.5).

We have proposed a new, simple method performing symbolic substitution. There is no need for any optical elements and imaging system, the LED's are used as both a light source and a control element in the operation. The research is significant for making integrated optical computing elements.

4. REFERENCES

[1] A. Huang, "Parallel Algorithms for Optical Digital Computers" , in Technical Digest, IEEE Tenth International Optical Computing Conference, (1983), pp13-17.

[2] K. Brenner, A. Huang, and N. Streibl, "Digital Optical Computing with Symbolic Substitution" , Appl.Opt.25.18. (1986).

[3] J. Tanida and Y. Ichioka, "Optical Logical Array Processor Using Shadowgrams" , J. Opt.Soc.Am.73.6.(1983).

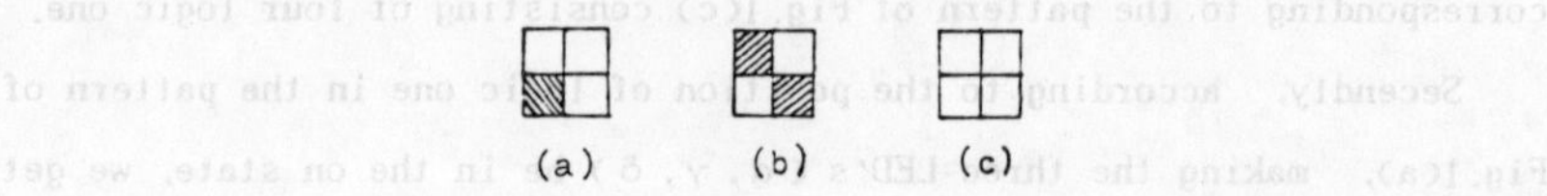

Fig.1. Patterns consisting of four cells: (a)recognition pattern; (b)substitution pattern; (c)pattern containing four logic one.

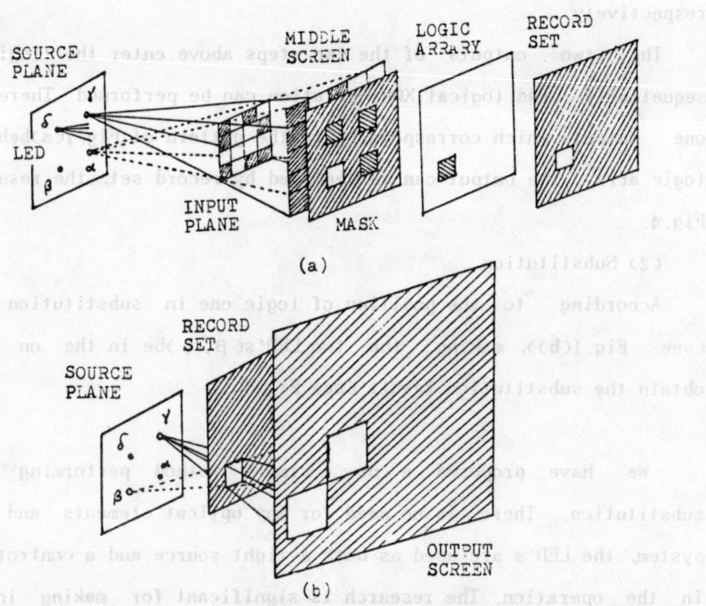

Fig.2. Schematic diagrams of implementing symbolic substitution using shadow-casting: (a)recognition phase; (b)substitution phase.

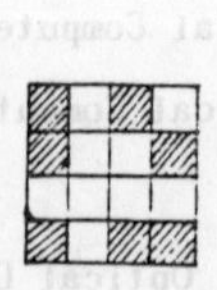

Fig.3. Input image

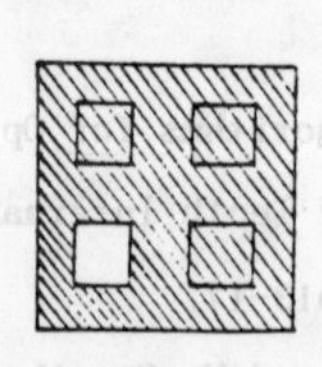

Fig.4. Record result after recognition

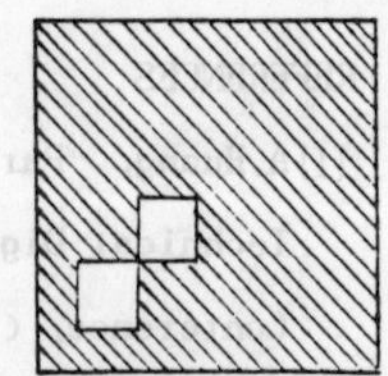

Fig.5. Result after substitution

A High Precision Fourier Transformer with Light Intensity & Polarization Encoding*

Qian Qiu-Ming Li Qing-Xiong
Liu Yao-he Wang Zhi-Jiang

Shanghai Institute of Optics and Fine Mechanics,
Academia sinica
P. O. Box 8216, Shanghai China

1. INTRODUCTION The optical two-dimension discrete Fourier transform (2-DFT) as a field of optical computing is quiet important. It can be implemented by coherent or incoherent optics. Coherent optical systems can compute the squared modulus of the 2-DFT simply and instantaneously,but it is limited by speckle noise and the proper spatial light modulators as input transducers. Though it is easy to show the amplitude and phase of the 2-DFT of a object or a pattern with the transform lens, the necessity of square-law detection makes difficult to separate the amplitude and phase components of the 2-DFT spectrum. In order to reduce background noise, incoherent optical Fourier transform systems have significant advantages over coherent systems. In the past several years, many systems have been proposed and realized.[1] But each of them has some disadvantages limiting its utility.

Here we present the principles and the implementation (including optical scheme)of a high precision whole parallel analog optical 2-DFT system with light intensity & polarization (LIP) encoding. In this system, the input data (real numbers) are encoded by LIP and nuclear mask of 2-DFT is constructed by 2N*N different squares whose optical transmissities and polarization angles correspond to the matrix

*This work was suported by Chinese Natural Science Fund

elements of 2-DFT nucleus. If the input data are encoded bit by bit before LIP encoding and the final outputdata are adjusted by microcomputer, the results of digital 2-DFT can also be obtained easily.

2. LIP ENCODING METHODS Any real number can be represented as follows:

$$M = |M| * [1+\mathrm{sign}(M)]/2*(+1) + |M| * [1-\mathrm{sign}(M)]/2*(-1)$$

Where

$$\mathrm{sign}(M) = \begin{cases} +1 & M>0 \\ 0 & M=0 \\ -1 & M<0 \end{cases}$$

In optics, it can be implemented by the following rules (i.e. LIP encoding):

A. Two independent polarization diretions (K_x, K_y) represent (+1,-1);

B. The light intensities of two independent polarization states, {I_x, I_y}, represent { $|M| * [1+\mathrm{sign}(M)]/2$, $|M| * [1-\mathrm{sign}(M)]/2$}, i.e:

$$I_x = |M| * [1+\mathrm{sign}(M)]/2$$

$$I_y = |M| * [1-\mathrm{sign}(M)]/2$$

Therefore $M = I_x * K_x + I_y * K_y$

Obviously, the following condition is always satisfied:

$$\min(|I_x|, |I_y|) = 0$$

3. THE CONSTRUCTION OF NUCLEAR MASK According to the theory of 1-DFT (2-DFT is equivalent to two times 1-DFT), We have

$$F_m = 1/N * \sum_{n=0}^{N-1} f_n [\exp(i2\pi/N)^{mn}]$$

$$m, n = 0, 1, \ldots, N-1$$

It may be represented by matrix form:

$$F = Cf + iSf$$

where $C_{mn} = 1/N * \cos(2\pi mn/N)$

$S_{mn} = 1/N * \sin(2\pi mn/N)$

When f is a real vector, the first term of above formula is the real part of 1-DFT spectrum and the second is the imaginary part. The nuclear elements of 1-DFT C_{mn} and S_{mn} can be expressed as follows:

$$C_{mn}=\text{sign}(C_{mn})*|C_{mn}|$$
$$S_{mn}=\text{sign}(S_{mn})*|S_{mn}|$$

In our optical 2-DFT systems, Let the transmissities of all mask squares correspond to $|C_{mn}|$ and $|S_{mn}|$ and the rotation angles Q_{uv} of them correspond with $\text{sign}(C_{mn})$ and $\text{sign}(S_{mn})$, where u,v=0,1,...,2N-1(When $\text{sign}(C_{mn})=+1, Q_{uv}=0^0$. When $\text{sign}(C_{mn})=-1, Q_{uv}=90^0$, similarly for S_{mn})

4. <u>SOME ELEMENTRAY CALCULATIONS ON 2-DFT AND THEIR IMPLEMENTATIONS IN OUR SYSTEM</u> There are two foundmental arithmetic operations in 2-DFT or 1-DFT.One is the multiplication $W_{mn}=f_n*C_{mn}$. Another is the summation of all W_{mn} with respect to n (For the convenience of the following discussion, we shall only describe formulas relating to C_{mn}. similarly for S_{mn}).

A. When the intensities and polarization directions of incident light array are corresponding to the input datum $|f_n|$ and $\text{sign}(f_n)$, the output light intensities from 2-DFT nuclear mask are equal to the values of W_{mn} (m, n=0, 1, ..., N-1), if the mask is constructed as above.

B. Using optical lens(including cylinder lens and spheric lens), with incoherent optics, it is easy to obtain the summation of W_{mn} with repect to n (n=0,1,..,N-1).

5. <u>IMPLEMENTATION OF OUR SYSTEM WITH INCOHERENT OPTICS</u>

Fig.1. shows the optical scheme of the high precision 2-DFT with light intensity and polarization encoding when N=2. First, the input datum encoded by LIP are transfered from light source array to input plane through optical fibers whose positions and directions in input plane are related to m, n. Second, all the divergent light rays become parallel when pass through microlens array. The centra and axis of every microlens can be determined by m, n. Third, all light rays from microlens array diverge in x- direction by cylinder lens, and then transit nuclear mask (In our experiment, the nuclear matrix is $(C\ S)^T$, because the formula F=Cf+iSf can be expressed as $(F_R\ F_I)^T=(C\ S)^T*f$.Finally the summations in Y-direction are produced by another cylinder lens. After passing through polarizing analyzer the output light intensity is received by detectors array and sent to microcomputer and divided into real and imaginary parts again by microcomputer.

Now we have find the results of 1-DFT.If above proceduces are repeated for real and imaginary parts of 1-DFT spectrum (Note: Here we must exchange rows with lines) and then put two new results together, we can obtain 2-DFT spectrum imediately.

6. DIGITAL INCOHERENT OPTICAL 2-DFT If the input data are encoded bit by bit before LIP encoding, we can make digital incoherent optical 2-DFT in our system. For example: When $f_{mn}=(f_{mn2}f_{mn1}f_{mn0})_m$, we have

$$F=(C+iS)(f_2f_1f_0)_m$$
$$=(F_2F_1F_0)_m$$

Where $F_k=(C+iS)f_k$ k=0,1,2. Obviously, the precision of 2-DFT can be improved by this methods. But it has some disadventages, such as the higher precision is obtained,the longer transform time is necessary.

7. CONCLUSION The 2-DFT system described above is complete parallel. The time of optical operation is about 0.1ns (e.g.2N*N additions and multiplications per 0.1ns). The system precision and operation speed are limited by the light sources and detectors though we can encode the datum one bit

by one bit.If the precision of detectors is 8-bits and scanning pixels are 32*32, the 2-DFT spectrum of a pattern (32*32 dots) can be produced in several us.

Reference:

1. Roger L.EAston,Jr.,A.J.Ticknor,and H.H.Barrett,"Two dimensionnal complex Fourier transform via the Radom transform" Appl.Opt.24,3817(1985)

FIG.1. Optical scheme

1. INPUT FIBERS
2. INPUT PLANE
3. MICROLENS ARRAY
4. CYLINDER LENS
5. NUCLEAR MASK OF 2-DFT
6. CYLINDER LENS
7. POLARIZATION ANALIZER
8. DETECTORS ARRAY

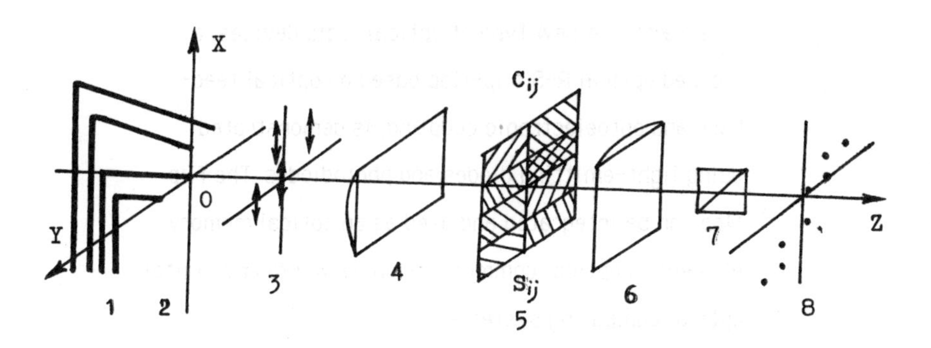

CLOCKED OPTICAL R-S FLIP-FLOP BASED ON OPTICAL FEEDBACK AND OPTOELECTRONIC COUPLING

Yan Shi, Kai-De Zha and Yi-Li Guo

Department of Information Electronics

Tsinghua University

Beijing, People's Republic of China

Indexing terms: Optoelectronics, Optical logic, Digital optical computing

Abstract: A new type of optical logic devices, a clocked optical R-S flip-flop based on optical feedback and optoelectronic coupling, is demonstrated using light-emitting diodes and photodiodes. The flip-flop can be integrated and used as an optical memory element in optical communication networks and digital optical computing systems.

Introduction: Optical flip-flops, which can perform memory functions, are fundamental elements of digital optical computing systems and basic components of many types of functional devices in optical communication networks. For the past few years, some types of optical flip-flops have been constructed. One type of the optical flip-flops was realized by combining optical logic gates, which were implemented by optical bistable devices, with complicated circuitry.[1] Another type of the optical flip-flops using polarization-bistable lasers was reported by J. M. Liu and Y. C. Chen,[2] which has to operate at a rather low temperature. In this letter, a new type of optical R-S flip-flop consisting of light-emitting diodes(LEDs) and photodiodes(PDs) is described, which has a very simple structure and has not any bistable devices.

Devices description and operation: The optical R-S flip-flop consists of two LEDs and five PDs, as shown in Fig. 1. The main part of the flip-flop is a combination of two linear devices based on optical feedback. The linear device is an electrical combination of a PD, a transistor and an LED, as shown in Fig.2(a), whose optical output power decreases as the optical input power increases. The two linear devices are connected by optical feedback to construct an unclocked optical R-S flip-flop, or basic optical R-S flip-flop, as shown in Fig.2(b). The optical output of LED_1 is fed optically to the input of PD_2 and the optical output of LED_2 to the input of PD_1. The stable state of the circuit is that one of the LEDs is at ON(light-emitting) state and the other at OFF(no emitted light) state. The output states of the two LEDs constitute the normal output, Q, and its complement, $\bar{Q}$, respectively. PD_3 and PD_4 are used to detect the optical input signals, R and S, respectively.

The process of the operation of the unclocked optical R-S flip-flop can be seen by the following consideration. Suppose the original states of the optical input signals, R and S, are both "0", the output states of the flip-flop will remain unchanged because the device is in stable state. If the optical input signal S becomes "1" and R is still "0", the current through the transistor T_1 will increase. It will then decraese the current through LED_1 and make the optical output of LED_1 decrease. Thus the optical output of LED_2 will increase due to the decrease of the current through T_2 caused by the decrease of the optical output of LED_1 and as a consequence the optical output of LED_1 will decrease still further. This cycle of events will repeat itself very quickly. Finally, LED_1 will turn off and LED_2 will turn on. That is, the optical output of the flip-flop, Q, will be "1". In reverse, if S is "0" and R is "1", LED_1 will turn on and LED_2 will turn off. That is, the optical output of the flip-flop, Q, will be "0". The condition that both S and R are "1" is forbidden in conventional electronic R-S flip-flop because it results in an indeterminate state. In our optical R-S flip-flop, if both S and R are "1", the optical output state will be that both Q and $\bar{Q}$ are "0" because of the structure of our flip-flop.

To get a clocked optical R-S flip-flop, PD_5 is introduced, as shown in Fig.1, to detect the optical clocked-pulse (OCP). The process of the operation of the clocked optical R-S flip-flop is similar to that of the unclocked optical R-S flip-flop. However, in the absence of the OCP, changes in logic state at the optical data input cause no change in the optical output of the flip-flop. At the moment when an OCP arrives, whether the optical output states change or not and what states they change to depend on the states of the optical input signals at the same moment. If both S and R are "0" when an OCP arrives, the optical output state will remain unchanged. If S is "1" and R is "0" when an OCP arrives, LED_1 will turn off and LED_2 will turn on. Thus the optical output Q will be "1". In reverse, if S is "0" and R is "1" when an OCP arrives, Q will be "0". If both S and R are "1" when an OCP arrives, which is forbidden in electronic R-S flip-flop, the optical outputs Q and $\bar{Q}$ will be both "0" at the moment when the OCP is present. Upon the OCP disappearing, one of the two optical outputs, Q and $\bar{Q}$, will be "1" and the other will be "0" at random.

Conclusion: A new type of clocked optical R-S flip-flop, based on optical feedback and optoelectronic coupling, has been successfully demonstrated using LEDs and PDs. By using higher speed optoelectronic elements, the flip-flop has a shorter swicth-on time than other types of optical flip-flops which are based on optical bistability. The structure of the flip-flop is very simple since it is neither a combination of optical logic gates nor based on optical bistability. The flip-flop consists only two LEDs, the minimum number for the construction of an optical flip-flop based on optical feedback and optoelectronic coupling. The number of PDs is also minimized. Hence the clocked optical R-S flip-flop can easily be hybridly integrated on a single chip. As an optical memory element, it can be expected to be used in digital optical computing systems and optical communication networks in the near future.

Acknowledgments: The authors would like to thank Dr. Min-Cai Xie for providing the photodiodes and Dr. Xiao-Lei Liu for helpful discussions. The work is supported by the National Nature Sicence Foundation of China.

References

1 OKUMURA, K., OGAWA, Y., ITO, H., and INABA, H.:'Optical bistability and monolithic logic functions based on bistable laser/light-emitting diodes', *IEEE J. Quantum Electron.*, 1985, QE-21, pp. 377-382

2 LIU, J. M., and CHEN, Y. C.: 'Digital optical signal processing with polarization-bistable semiconductor lasers', *ibid.*, 1985, QE-21, pp.298-306

LEGENDS TO ILLUSTRATIONS

Fig.1 *The clocked optical R-S flip-flop*

a the circuit of the flip-flop

b the operation of the flip-flop

c the logic diagram and characteric table of the R-S flip-flop

Fig.2 *The unclocked optical R-S flip-flop*

a the electrical combination of a PD, a transistor and an LED

b the circuit of the flip-flop

c the operation of the flip-flop

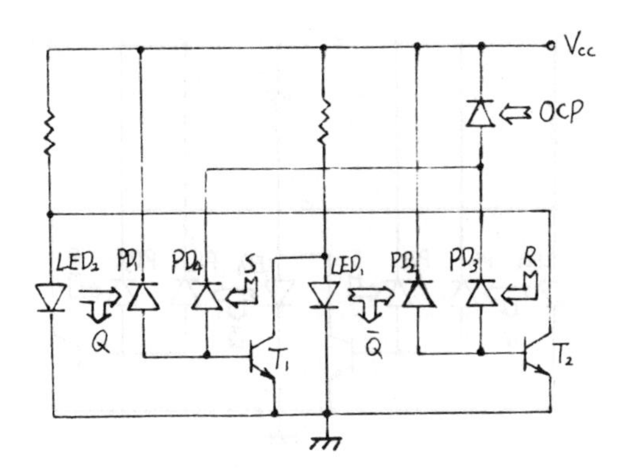

a

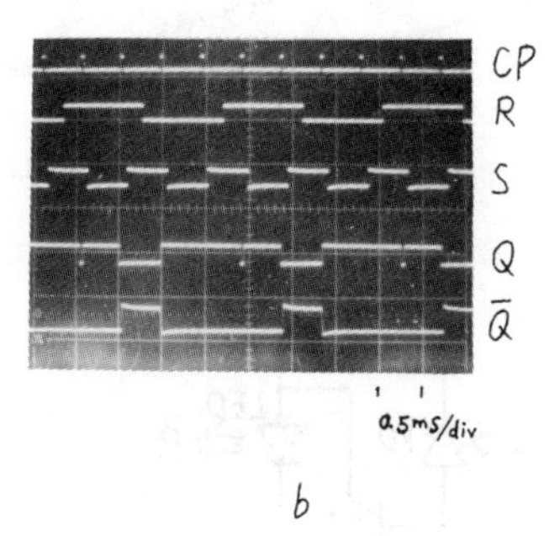

b

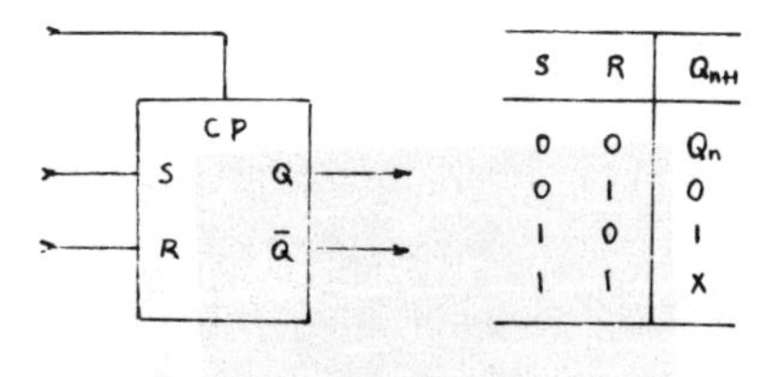

S	R	Q_{n+1}
0	0	Q_n
0	1	0
1	0	1
1	1	X

c

Fig. 1

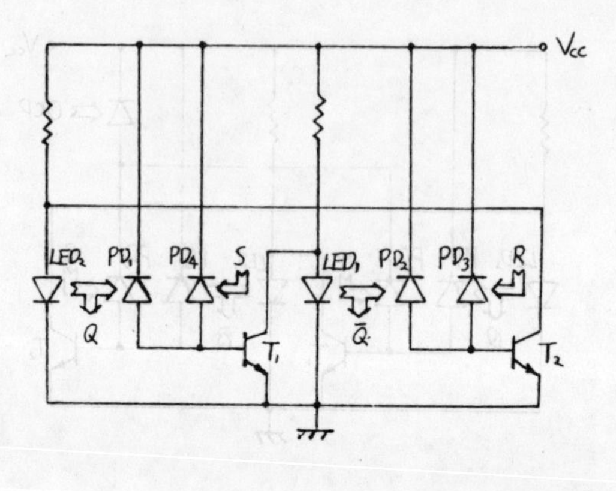

a

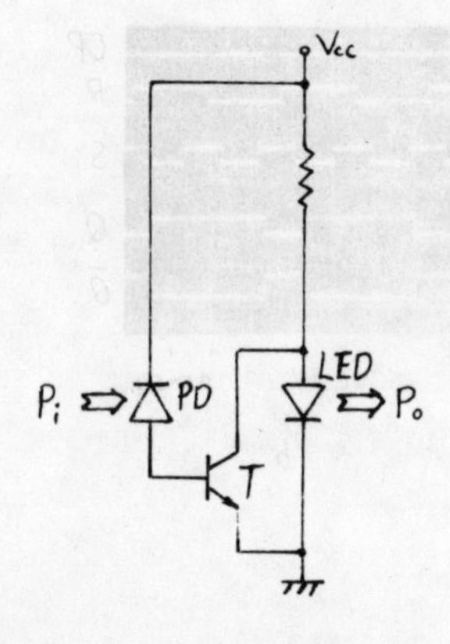

b

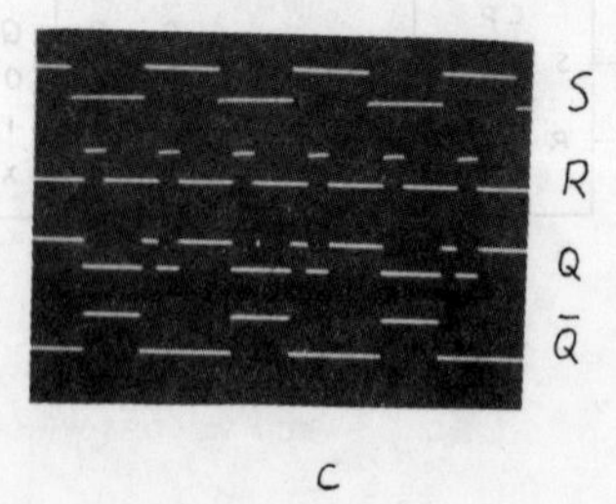

c

Fig. 2

A NEW OPTICAL ADDER

Weng Zhaoheng
Changchun Institute of Optics and Fine Mechanics, China
Chen Yongnian, Lu Yulin
Shanghai University of Science and Technology, China

ABSTRACT

A new optical adder is proposed in this paper. It is constructed by making use of the N-shaped mutual-compensative characteristic of the reflection and transmission output of F-P type's bistable optical elements. This type of adders have a very simple structure and a very short delay time.

I. INTRODUCTION

Operational unit is a very important part in computers. Adder is the main component of it. In circuits based on electronic logic devices, full-adders are usually constructed with half-adders which are constructed with logic gates, while logic gates are constructed with nonlinear electronic components. The delay time of a full-adder is usually 4-6 levels gate-delay. To form an n-bit adder, n full-adders are needed, and its delay time is n times of that of a full-adder. If parallel carries are used, the delay time may be reduced, but the logic gates needed will increase a large number.

The commom way of constructing adders using optical devices is just the same as using electronic circuits, that is, first, constructing optical logic gates with nonlinear optical devices, and then constructing adders with logic gates. In this way, a lot of optical components will be needed and the delay time will be very long.

Optical devices have many special characteristics other than electronic devices. We can completely avoid the complex structure of electronic circuits if they are suitablly used.

A new type of optical adder is proposed in this paper. The N-shaped mutual-compensative characteristic of the reflection and transmission output of F-P type's bistable optical elements is used. Using such characteristic, only 3 elements are needed for constructing a full-adder. The delay time of it is only the sum of that of two elements. Its structure is very simple and it is easy to be realized. When this type of adders are used to construct n-bit adders, the parallel carries are not needed because of the short delay time of each full-adder. Thus, the whole structure become more simple.

II. TRADITIONAL STRUCTURE OF ADDERS

The traditional way of constructing adders is to construct full-adders using logic gates constructed by nonlinear electronic elements first, and then to construct adders using such full-adders. Fig. 1 shows two typical structure of full-adders constructed by NAND gates. Where Ai and Bi are two input signals, C_{i-1} is the carry come from the lower bit, Si is the sum and Ci is the carry. The delay time of these two circuits are 4 and 6 levels gate-delay respectively.

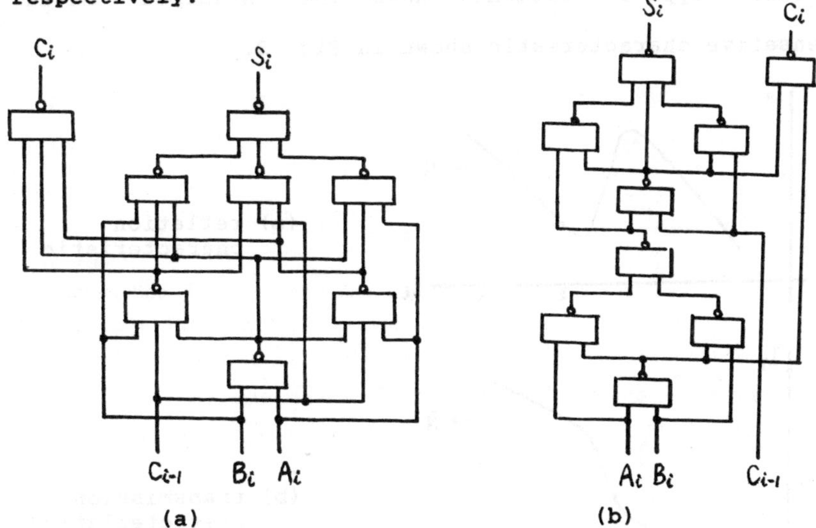

Fig. 1 Two typical structure of full-adders

We may connect such full-adders in series to form an adder, but it results in a very long delay time (about bits*3 levels gate-delay) because of the series characteristic of carries. In practical, grouped parallel carries method (commonly 4 bits a group) is usually adopted. In such case, the delay time may decrease, but more a lot of logic gates will be needed and thus make the circuit more complex.

III. THE CHARACTERISTIC OF BISTABLE OPTICAL ELEMENTS

We find that the structure of optical adders may be simplified by using the special characteristic of some nonlinear optical elements. A particular nonlinear optical element——F-P type's bistable optical element is used for constructing adders. The reflection and transmission output of this type's elements have the N-shaped mutual compensative characteristic shown in Fig. 2.

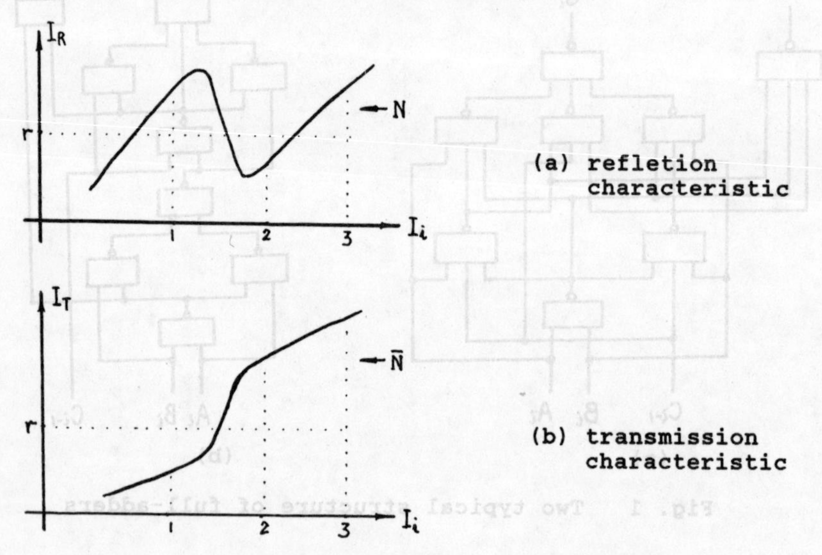

Fig. 2 The N-shaped mutual compensative characteristic of F-P type's bistable optical elements

The input beams are all of unit intensity. The output is regarded as level '1' when the intensity of output beam is greater than r, otherwise it is regarded as level '0'. Thus, according to the reflection characteristic, the output I_R will be '0' if there is no input beam or there are two input beams, otherwise I_R will be '1'. According to the transmission characteristic, the output I_T will be '1' when the intensity of input signals is not less than 2, otherwise I_T will be '0'.

In order to get a steady and uniform output optical level, the saturation characteristic of bistable optical elements may be used for output shaping. The saturation characteristic of bistable optical elements is shown in Fig. 3. It can be ajusted such that the output optical level will be '0' when the input is less than r and it will be '1' when the input is greater than r.

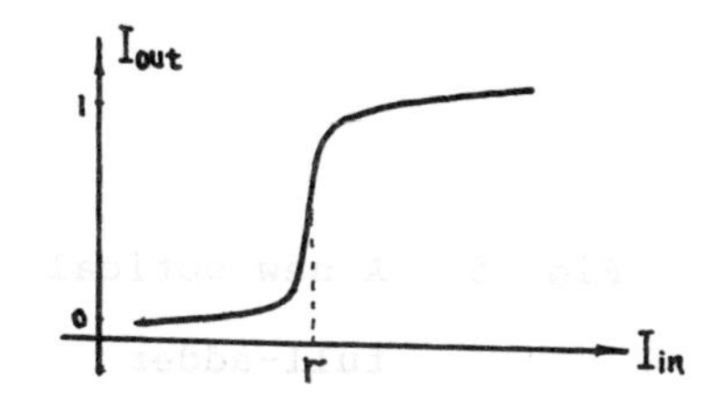

Fig. 3 The saturation characteristic of bistable optical elements

IV. A NEW OPTICAL ADDER

A new type of optical adders can be constructed using the characteristic of the F-P type's bistable optical elements described above.

The truth table of full-adders is shown in Fig. 4. Si will be '1' when the number of '1' at input is odd. Ci will be '1' when there are two or more '1' at input.

Ai	Bi	Ci-1	Si	Ci
0	0	0	0	0
0	0	1	1	0
0	1	0	1	0
0	1	1	0	1
1	0	0	1	0
1	0	1	0	1
1	1	0	0	1
1	1	1	1	1

Fig. 4 The truth table of full-adders

From the N-shaped mutual compensative characteristic of F-P type's bistable optical elements, It can be seen that I_R

is just corresponding to the sum Si while I_T is just corresponding to the carry Ci. Thus, one F-P type's bistable optical element and two bistable optical elements used for output shaping can form an optical full-adder. Fig. 5 shows such a full-adder, where N and $\bar{N}$ represents a N-shaped mutual compensative logic device, Sa_1 and Sa_2 are two same saturation bistable switch devices used for output shaping.

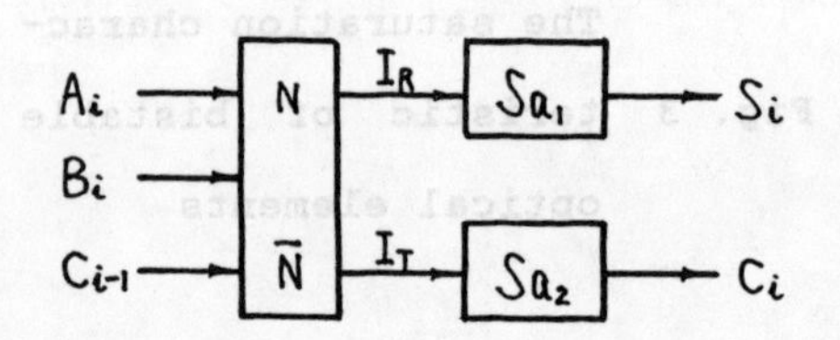

Fig. 5 A new optical full-adder

When such full-adders are used to construct adders, the parallel carries used in traditional adders for decreasing the delay time of carries will not be needed because of the very short delay time of each full-adder. Connecting the full-adders simply in series will be a practical way. Such structure of adders is shown in Fig. 6. The delay time of this adder is the sum of that of 2*n elements.

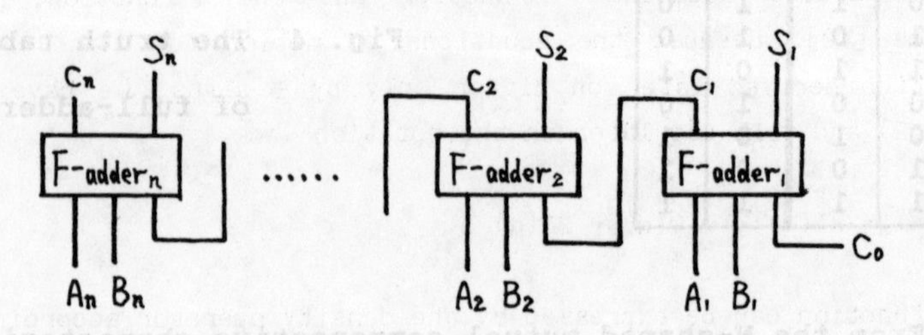

Fig. 6 The structure of n-bit adders

V. CONCLUSION

A new optical adder is proposed in this paper. Because the special characteristic of bistable optical elements is used, a full-adder consists of only three elements, and the delay time is also only the sum of that of two elements. Compared with traditional adders, it has the advantage of simple structure and short delay time. So it will be an optical adder of bright prospect.

REFERENCES

[1] Special issue on optical computing, Proc. of IEEE, July, 1985

[2] Xu Zhenyu, Priciples of electronic digit computers, Tianjin University, 1979

[3] Lu Yulin, A new optical sorting unit OSU-3, Proc. of National Computer Youth Conf.'87, Harbin, Aug. 1987

FUNCTION ANALYSIS
A GOOD APPROACH TO OPTICAL COMPUTING SYSTEM

Liang Minhua, Wu Shudong, Wang Zhijiang
Zhang Xiang, Li Qingxiong

Shanghai Institute of Optics and Fine Mechanics,
Academia, Sinica
P.O.Box 8216, Shanghai, China

ABSTRACT: The idea of function analysis, a useful method for optical computing system, has been proposed here. Three essential conditions for optical computing system are analysed and a quality factor of an optical computing system is proposed. The relations among computing speed, function, architecture and space are discussed. The examples of applying the method are given too.

To date many significant researches, such as symbolic substitution[1], look-up processing[2], associative memory[3], [4], and matrix processing [5] etc., have been made for optical computing architectures and algorithms, but most of those architectures and algorithms still want reliable and practical hardware systems. How to design effective optical computing systems according to the means that optics and electronics can offer is of great significance. Function analysis can help to find out these effective systems.

Function analysis implies that a computing system can be regarded as a function set which is constituted by a group of functions and elements in the set can be decomposed into sets that are constituted by one-level-lower functions, while each element in a set corresponds to at least one physical structure form, therefore researches and designs about system's functions can be done through studying spatial distribution and interrelation of those functional elements.Functions indicate the architectures and algorithms of a computing system.

Finishing any tasks needs a process which can be considered as a function set that is made up of a group of functions.If a task's function set Ft $\subseteq$ a system's function set Fs,this system can finish the task,i.e. realizing the process of the task.If Fs $\subset$ Ft,only a portion of the task can be done by the system.So a function set can be used to represent a system,a task or a process,therefore the designing of an optical computing system is to find out a system with such a Fs that it includes some tasks,i.e.Fs $\supseteq$ Ft_i (i=1,2..N)

According to the viewpoint of function analysis,a computer system must have some basic functions.They are (1) programmable and automatical operating; (2)information processing(including numeral calculating,data processing,logical inference,intelligent processing); (3)man-machine information exchange.

A system that have a calculating function must meet two essential conditions,one is that there are some independent cardinal numbers ruled by a certain numeral system in an abstrac form,and the other is that in the domain of specific physical form each cardinal number have to correspond to one physical structure form.Those cardinal numbers have such a property that they can constitute any elements in a certain numeral domain and can realize each kind of element-between transforms in that domain by means of operating symbols.A group numbers with the property mentioned above is called perfect or complete in that domain.Such a group of numbers is not sole.Similarly the cardinal physical structure forms have such a property that those forms can represent all elements and realize each kind of element-between transforms in the physical domain through sequential arranging and combining themselvs in space.

Being similar with a computing system,a system with a function of logical inference must have two essential conditions of its own,(1)some independent cardinal logical elements ruled by logic rules in the domain of abstract theoretical form,(2) in the domain of specific physical form each cardinal element having one physical structure form to represent itself.The cardinal logical elements and the cardinal logical structure forms have similar properties with the cardinal numbers and the cardinal numeral structure forms.

If some theoretical forms corresponding to some cardinal logical elements are found out in a certain numeral domain and the functions of the cardinal logic structure forms can be realized by the cardinal structure forms in that numeral domain through spatially sequential arranging and combining of the numeral structure forms,the corresponding computer system is perfect or complete both in this numeral domain and in the domain of logic function,and vice versa.

In addition to the essential conditions stated above,another essential condition,i.e. information carrier,is needed in a computing system which has either computing function,or logic function,or both.From level of the viewpoint of function analysis,the higher the functions of a computing system are,the more complicated the structures are,and in addition, the larger the volume is.Functional elements in a system's function set are connected in space according to the priciple of function continuity.

In an optical computing system,the computing speed has such a following interrelations with the speed of the cardinal structure Vc,channel's number Nch,length of the channel Lch,capacity of channel Cch and capacity of information carrier Cic,that is

$$Vcomp \propto Vc, Nch, Cch, Cic, 1/Lch$$

the quality factor of an optical computing system can be evaluated by

$$Q = \frac{[FUNCTION]*[Vcomp]}{[STRUCTURE]*[VOLUME]*[CONSUMED\ ENERGY]}$$

It is important for an optical computing system to choose what kind of functions utilizing optical inherent parallelism to the utmost other than to have functions of computing and logic.

Using the approach of function analysis we have designed two acousto-optical digital multpliers(serial and parallel) and corresponding digital vector-multplication systems.We shall look for better optical computing systems by means of the above approach.

This work was supported by Chinese Natural Science Fund.

References

[1] K.H.Brenner,A.Huang,N.Streibl,'Digital optical computing with symbolic substitution' Appl.Opt.25,3054 (1986)

[2] T.K.Gaylord and M.M.Mirsalehi,'Truth-table look-up processing:number representation,multilevel coding,and

logical minimization' Opt.Engin.25,22(1986)

[3] B.Macukow and H.H.Arsenault,'Optical associative memory model based on neural networks having variable interconnection weights' Appl.Opt.26,924(1987)

[4] G.J.Dunning,et al,'All-optical associative memory with shift invariance and multiple-image recall' Opt.Lett.12,346(1987)

[5] D.Casasent and J.Jackson,'Laboratory optical linear algebra processor for optical control' Opt.Comm.60,1(1986)

FUNDAMENTAL PARALLELISM IN OPTICS

Liang Minhua, Wu Shudong, Wang Zhijiang
Zhang Xiang, Li Qingxiong

Shanghai Institute of Optics and Fine Mechanics,
Academia, Sinica
P.O.Box 8216, Shanghai, China

ABSTRACT: In order to utilize the optical computing to the best, the fundamental parallelism in optics is discussed in detail. The measurement of the parallelism is proposed. The parallelity has some relations with encoding. The wider the word width for encoding is, the lower the parallelity of the system is. The inherent parallelism of optics appears in the optical propagation in free space, which has the capability of pipe-line processing. Resource-sharing optical computing system can ease the problem of bandwidth matching.

Most of the schemes of optical computing architectures and algorithms are based on the optical parallelism[1],[2],[3],[4]. Until now not enough attentions have been given to the types of optical parallelism, the forms of existance, the way of measuring it and the other characteristics yet. The solution of these problems are very important for the research and design of optical computing architectures and algorithms which can utilize the parallelism of optics to the best.

Generally speaking, parallelism has two meanings, simultaneity and concurrency[5]. The former implies that two or more events happen simutaneously and the latter implies that two or more events happen at the same period of time. Parallelism can be realized in the following ways, 1) time-interleaving, i.e. many courses of processing are staggered in time and use each part of the same set of hardwares in turn and in a way of temporal-overlap;2)resource-replication,i.e.similar resources are deployed repeatedly;3)resource-sharing,i.e. many users alternately use the same set of hardwares in a certain order.According to the parallel level there can be five approaches:memory operating parallelism,processor's operating-step parallelism,processor operating parallelism,instruction parallelism and task parallelism.

The fundamental parallelism in optics means the sum of independent channels that optical system can give as the optical waves and information carriers propagating in the system.In a computing system it is operants that are actually processed.Those operants are coded in a certain form and each cardinal coding element occupies one channel.Being regarded as the measurement for the parallelism of an optical computing system,number of the coded operants processed in a time unit reflects a real data-processing level that a computing system has.The channel number Nc word width W and the parallelity of a system are related in the following way:Nc= $\sum a_i W_i$; $\sum a_i$ =N. N is the measurement of parallelity,Wm the maximum word width and a the number of the operants which have W word width.With the sum of a system's channels given ,the system's parallelity decreases as the channels used for encoding increase.

The dimensions that can provide channels are space(1-D, 2-D or 3-D),time and frequency(spatial and temporal).The information carrier can be either intensity (or amplitude),or intensity+polarizing state(or amplitude+polarizing state),or intensity+phase (or amplitude+phase).As regarding to optical information,there can be two basic encoding approaches,one is time-expansion-encoding,i.e. encoding by means of the interrelations among the information carriers distributed over time-dimension,the other is space-expansion-encoding,i.e. encoding by means of interrelations among the information carriers distributed over space-dimensiion.With Nc given,a system has a constant parallelity as for the same encoding word width of both time-expansion-encoding and space-expansion-encoding.

Under a free space propagating conditon Nc that the monochromatic wave can provide is equal to the value of optical spatial effective size devided by the system's spatial resolution.Usually this value is very large.According to time-interleaving,let each function device be specialized and optical wave pass through these devices in a certain order,and this will constitute a pipe-line processing.Such a typical optical system is a two-lens system which can make two Fourier-transforms[6].Generally most of the optical systems have the inherent capability of pipe-line processing and that is just the place the optical massive parallelism lies. It is imfortunate that most of them appears naturally in the form of analog processing.Hence,how to utilize such an inherent capability to do digital calculations is one of the very important problems.In such a system another problem is bandwidth matching i.e. equality of bandwidth among the components of the system.

If the optical wave is constrained in space,i.e. in optical fiber or optical wave guide,and appears in the form of pulses to convey the informations,or if the wave is in the pulse format in free space,a monochromatic wave can just have only one channel.Intensity is commenly used as information carrier.Under this circumstance the optical inherent parallelisn disappears.If those similar channels are arranged repeatedly(the lowest resource-replication),the channel's number increases ,for example,Tse computer system[7].

So far there're still no optical computing systems that utilize resource-sharing to boost system's parallelism.Based on current optical and optoelectronic technology,however, such systems can be easily worked out and furthermore the resource-sharing optical computing system can ease system's bandwidth matching problem.

In the viewpoint of parallel hierarchy,optics can have the following inherent parallelism:memory operating parallelism,e.g. holographical memory etc.;processor's operating-step parallelism,e.g. optical pipe-line processing system; processor operating parallelism,e.g. acousto-optical multipliers.

The fundamental parallelism in optics is discussed in the above context.As a conclusion,the optical inherent parallelism mainly shows the capability of pipe-line processing.A system's parallelity is related to encoding.The wider the word width used for encoding is,the lower the parallelity of a system is.Finally,the bandwidth matching problem can be eased in a resourse-sharing optical computing system.How to design optical computing architectures and algorithms based on the inherent pipe-line processing capability and other parallelisms is a subject which warrants further researches.

This work was supported by Chinese Natural Science Fund.

References

[1] A.Huang,'Architectural considerations involved in the design of an optical digital computer' Pro.IEEE 72,780(1984)

[2] T.K.Gaylord and M.M.Mirsalehi,'Truth-table look-up processing:number representation,multilevel coding,and logical minimization' Opt.Engin. 25,22(1986)

[3] G.J.Dunning,et al,'All-optical associative memory with shift invariance and multiple-image recall' Opt.Lett. 12,346(1987)

[4] D.Casasent and J.Jackson.'Laboratory optical linear algebra processor for optical control' Opt.Comm.60,1(1986)

[5]P.H.Enslow,Jr.(ed.),Multiprocessors and parallel processing,Jhon Wiley&Sons,1974

[6]S.H.Lee(ed.),Optical informatin processing fundamentals,Springer-Verlag Berlin Heidelberg New York 1981

[7] D.H.Schaffer and J.D. Strong,'Tse computers' Pro.IEEE 65,129(1977)

SQUEEZED STATE GENERATION FROM AN OPTICAL PARAMETRIC OSCILLATOR AND ITS APPLICATION TO INTERFEROMETRY

Ling-An Wu*, Min Xiao and H.J.Kimble

Univ. of Texas at Austin, Austin, TX 78712, U.S.A.

Squeezed states of the electromagnetic field have been generated by degenerate parametric down conversion in a sub-threshold optical parametric oscillator. Reductions in photo-current noise of up to 63% relative to the vacuum noise level have been observed in a balanced homodyne detector. A quantitative comparison with theory indicates that the observed noise reduction results from a field that in the absence of linear losses would be squeezed more than tenfold. From the experimental data an explicit demonstration of the Heisenberg uncertainty principle is obtained, indicating that the field produced by the down-conversion process is a state of minimum uncertainty.

By employing this squeezed light in a Mach-Zehnder set-up to measure the phase modulation in propagation along the interferometer arms, an improvement of 3.0dB in signal-to-noise ratio relative to the shot-noise limit has been obtained. This increase in sensitivity is currently limited by losses in propagation and detection and not by the degree of available squeezing.

*On leave from the Institute of Physics, Academia Sinica, China.

SECOND HARMONIC GENERATION IN A MULTI-MODE LASER CAVITY

Xiao-Guang WU

Université Libre de Bruxelles, Campus Plaine C.P. 231
1050 Bruxelles, Belgium

ABSTRACT

We analyze a theoretical model of SHG(second harmoni generation) which compared satifactorily with recent experimental observations. This model is characterized by mode competition and mode inhibition.

1. INTRODUCTION

SHG has become an actively studied field of nonlinear optics since the advent of lasers. The first experimental evidence of SHG was reported in 1961 by Franken et al [1] and since then a number of theoretical analyses has been reported (see e.g. Ref. 2-5). Very recently, the experimental realization of SHG in a multi-mode laser cavity was reported by Baer [6,7] who compared the experimental results with a simple theoretical model [8] (rate equation approximation). Baer's work is of much impotance as it indicates the nature of the mechanisms that dominate, in at least one experimental set-up, the phenomenon of SHG in a laser cavity.

Baer's model is a set of rate equations for the fundamental intensities and gain functions with all the second harmonic modes adiabatically eliminated. For simplisity, we use scaled variales and parameters, so that Baer's model can be written [9]

$$
\begin{aligned}
dI_j/dt &= D_j - k_j - g\Big(I_j^2 + 2I_j \sum_{i\neq j}^{N} I_i\Big) \\
dD_j/dt &= D_{j0} - D_j\Big(1 + I_j + \sum_{i\neq j}^{N} b_{ij} I_i\Big)
\end{aligned}
\tag{1.1}
$$

where N is the number of fundamental modes, I_j the intensity of mode j with relaxation rate k_j, D_j the gain function of mode j with unsaturated value D_{j0}, g the nonlinear coupling coefficient, b_{ij} the cross saturation parameter between modes i and j and the time variable t has been normalized to the relaxation rate of the gain function. In Eqs.(1.1) the term $-gI_j^2$ represents the loss (for the fundamental mode j) due to the frequency doubling process whereas the terms $-2I_iI_j$ with $i \neq j$ represent losses due to the sum-frequency generation. As in our previous study on SHG in a single-mode laser cavity[4,5], we are interested in defining analytically the temporal nature of the output intensities. This requires first a determination of the steady states of Eqs.(1.1). A linear stability analysis of these steady states will determine domains of coexistence between stable steady-state solutions and Hopf bifurcations[10] where the steady-state solution looses its stability for a periodic output.

2. TWO-MODE OPERATION

when etalons placed in the cavity allow only two modes to ocsillate, Eqs.(1.1) are reduced to a set of four coupled equations:

$$
\begin{aligned}
dI_1/dt &= (D_1-k_1-gI_1-2gI_2)I_1 \\
dD_1/dt &= D_{10}-(I_1+bI_2+1)D_1 \\
dI_2/dt &= (D_2-k_2-gI_2-2gI_1)I_2 \\
dD_2/dt &= D_{20}-(I_2+bI_1+1)D_2
\end{aligned}
\qquad (2.1)
$$

where $b_{12}=b_{21}=b$.

2-1. Steady-State Solution

A steady state solution is defined by the condition $dI_j/dt=$ $=dD_j/dt=0$. Solving Eqs.(2.1) for steady state leads to three possible classes of solutions:

(1) The trivial solution

$$I_j = 0, \quad D_j = D_{j0} \; (j=1,2) \qquad (2.2)$$

It corresponds to the case where both modes are below the ocsillation threshold.

(2) The single-mode solution

I_1 can be zero and nonzero, and, independently, I_2 can be zero or nonzero. Single mode operation means that one mode (say I_1) is above the ocsillation threshold and the other mode (I_2) is below threshold:

$$I_1 > 0, \quad I_2 = 0 \qquad (2.3)$$

where $\quad I_1 = I_{1+}$

$$I_{j+} = \frac{1}{2g}\left(-(k_j+g)+\sqrt{(k_j-g)^2+4gD_{j0}}\right) \qquad (2.4)$$

The positivity of the intensity gives a condition on D_{j0}:

$$D_{j0} > k_j \qquad (2.5)$$

The critical value $D_{j0} = k_j$ is the oscillation threshold for mode j.

(3) The two-mode solution

The steady states for two-mode operation are the positive solutions of the following coupled equations:

$$\begin{aligned} D_{10} &= (I_1+bI_2+1)(k_1+gI_1+2gI_2) \\ D_{20} &= (I_2+bI_1+1)(k_2+gI_2+2gI_1) \end{aligned} \qquad (2.6)$$

Eqs.(2.6) lead to a quartic for the intensities. In two limit cases: (i)$g \ll k_j$ and (ii)$g \gg k_j$ we may express the intensities in a power series of a small parameter s defined by $s = g/(k_1k_2)^{1/2}$ for case (i) or by $s = (k_1k_2)^{1/2}/g$ for case (ii):

$$I_1 = \sum_{n=0}^{\infty} s^n I_{1,n}, \quad I_2 = \sum_{n=0}^{\infty} s^n I_{2,n} \qquad (2.7)$$

where the coefficients $I_{j,n}$ are defined in Ref.9.

2-2. Linear Stability Analysis

Here we give some results of the linear stability of the steady states of Eqs.(2.1). The detailed calculations can be found in Ref.9.

(1) Stability of the trivial solution

The linear stability analysis for the trivial solution (2.2) is easily performed in the usual way[10] and leads to the result that the trivial solution is stable if and only if $D_{j0} < k_j$. The corresponding domain is labelled A on Figs.1 and 2.

(2) Stability of the single-mode solution

The linear stability analysis of the solution (2.3) leads to the result that the solution $I_1 = I_{1+}$, $I_2 = 0$ is stable when $D_{20} < D_{20}^+$ where

$$D_{j0}^+ = (1 + bI_{i+})(k_j + 2gI_{i+}) \quad (i,j=1,2;\ i \neq j) \quad (2.8)$$

The corresponding domain is labelled C on Figs.1 and 2. $D_{20} = D_{20}^+$ is the oscillation threshold for mode 2. Furthermore we have $D_{20}^+ > k_2$, which means that the threshold for mode 2 becomes higher when the mode 1 oscillates. In other words the presence of mode 1 inhibits mode 2 when $k_2 < D_{20} < D_{20}^+$, because part of the gain available for I_2 in the absence of I_1 contributes to mode 1. That is, I_1 partly saturates the laser medium, thereby reducing the gain for I_2. If D_{20} is sufficiently large ($D_{20} > D_{20}^+$) to overcome the competition from I_1, I_2 builds up. Up to now we know that the single-mode operation solution

$$I_1 = I_1^+, \quad I_2 = 0 \quad (2.9)$$

is stable when $D_{10} > k_1$ and $D_{20} < D_{20}^+$. Similarly the solution

$$I_1 = 0, \quad I_2 = I_2^+ \quad (2.10)$$

is stable when $D_{20} > k_2$ and $D_{10} < D_{10}^+$. The corresponding domain is labelled B in Figs. 1 and 2. An interesting phenomenon is that the region $D_{10} > k_1$, $D_{20} < D_{20}^+$ and $D_{20} > k_2$, $D_{10} < D_{10}^+$ overlap in the parameter plane (D_{10}, D_{20}) and define a domain (labelled by E in Fig.2) where both solution (2.9) and solution (2.10) are stable. The particular state reached will depend on the initial conditions for Eqs.(2.1). As a result, mode 1

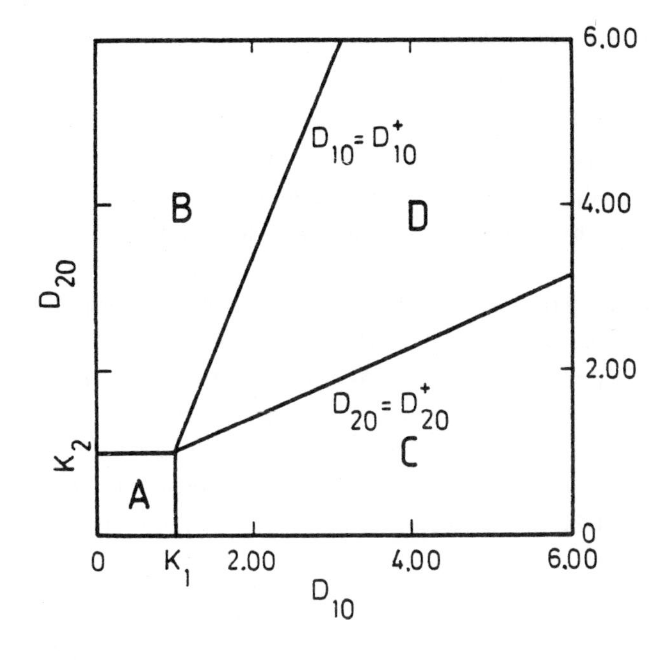

Fig.1 Four domains A, B, C and D in the parameter plane (D_{10},D_{20}) for $k_1=k_2=1$, $g=0.1$ and $b=0.25$. In domain A only the trivial solution is stable. In domain B(C) I_2 (I_1) builds up and takes the steady state $I_{2+}(I_{1+})$ while $I_1(I_2)$ remains zero. In domain D all the two modes start to ocsillate and reach a steady state.

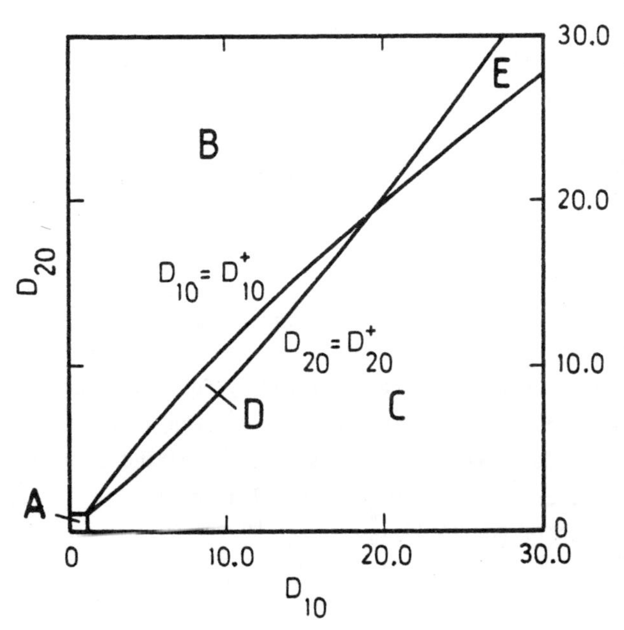

Fig.2 Five domains in the (D_{10},D_{20}) plane for $k_1=k_2=1$, $g=3.9$ and $b=0.25$. The stable states reached for I_1 and I_2 in domains A, B and C are the same as in Fig.1 while in domain D the two intensities I_1 and I_2 are time-periodic. Domain E is a bista ble domain where both two states $I_1=I_{1+}$, $I_2=0$ and $I_1=0$, $I_2=I_{2+}$ are stable.

can inhibit entirely mode 2 even when $D_{10} < D_{20}$.

(3) Stability of the two-mode solution

The stability of the solution $I_1>0$, $I_2>0$ is somewhat more difficult to assess. Here we only consider three special cases.

(3-1) $g<<1$, $k_j \sim 1$. The steady-state solution can be expressed by (2.7) and it is easy to verify that this solution is stable when both modes are above treshold.

(3-2) $g>>1$, $k_j \sim 1$. No steady state $I_{1s} > 0$, $I_{2s} > 0$ exists. The mode competition is so strong that one mode can inhibit entirely the other one.

(3-3) $k_1 = k_2$, $D_{10} = D_{20}$. This is a symmetric case where all the parameters are equivalent for mode 1 and mode 2 and therefore we seek symmetric solutions $I_{1s} = I_{2s}$. In section 3 we shall discuss the case of N-mode operation (N arbitrary) and the stability analysis will be performed only for the symmetric case. For N=2, the symmetric steady state is stable in the range $0 < I_{js} < I_H$ where

$$I_H = 1/(g-b-1) \qquad (2.11)$$

is a Hopf bifurcation which exists if and only if $g > b+1$. When $g < b+1$ there is no instability for the symmetric state.

When $g \sim k_j$ the linear stability of the two-mode steady state can only be analyzed numerically. An example of stability domains is shown in Ref.9.

To sum up, we have shown that a Hopf bifurcation can occur only when the two modes are oscillating. Otherwise only steady bifurcations will occur, corresponding to a transition from one oscillating mode to two oscillating modes.

3. N-MODE OPERATION

SHG in a N-mode laser cavity can be described by Eqs.(1.1). Because of the great number of variables, the analysis of such a system becomes rather elaborate. Hence we only discuss some restricted domains of parameters. In particular we assume $b_{ij} = b$ for all i and j.

The trivial solution of Eqs.(1.1) is

$$I_j = 0, \quad D_j = D_{j0} \; (j=1,2,\ldots,N) \tag{3.1}$$

which is stable if and only if $D_{j0} < k_j$. The nontrivial steady state is

$$I_{js} > 0 \; (j=1,2,\ldots,N) \tag{3.2}$$

The linear stability analysis of the solution (3.2) yields an Nth order characteristic equation which may be solved for arbitrary N in the symmetric case ($k_1 = k_2 = \ldots = k$, $I_{1s} = I_{2s} = \ldots = I$) and leads to the following results (see Ref.9 for detailed calculations):

(1) When $k < (2N-1)/b(N-1)$ and $(1-b)k < g < (2N-1)(1-b)/b(N-1)$ the bifurcation form the trivial solution is subcritical, indicating a domain of bistability. The finite intensity solution has two branches. The lower branch is unstable whereas the upper branch is stable from the limit point (corresponding to $I=I_1$) up to a Hopf bifurcation (corresponding to $I=I_H$) with

$$I_1 = \frac{g-(1-b)k}{(2N-1)(1-b)-g(N-1)b} \tag{3.3}$$

$$I_H = \frac{1}{g-(N-1)b-1} \tag{3.4}$$

(2) When $k > (2N-1)/(N-1)b$ and $\frac{(2N-1)(1-b)}{(N-1)b} < g < (1-b)k$ there are still two critical points I_1' and I_H where

$$I_1' = \frac{(1-b)k-g}{(2N-1)(1-b)-g(N-1)b} \tag{3.5}$$

and I_H is still given by (3.4). If $I_1' > I_H$, the bifurcation from the trivial solution is supercritical and the finite intensity solution is stable up to the first bifurcation I_H. If $I_1' < I_H$, the bifurcation from the trivial solution is subcritical and all finite steady intensities are unstable.

(3) When $g < \mathrm{Min}\left\{\frac{(2N-1)(1-b)}{(N-1)b}, (1-b)k, 1+(N-1)b\right\}$ the solution (3.2) is

is always stable for $I > 0$.

(4) When $1+(N-1)b < g < \text{Min}\left\{\frac{(2N-1)(1-b)}{(N-1)\,b}, (1-b)k\right\}$ the steady symmetric solution is stable up to $I = I_H$.

(5) When $g > \text{Max}\left\{\frac{(2N-1)(1-b)}{(N-1)b}, (1-b), 1+(N-1)b\right\}$ the solution (3.2) is always unstable.

The general case where all k_j and I_j are different cannot be studied analytically. It requires a systematic numerical integration of Eqs.(1.1).

4. NUMERICAL RESULTS

Here we present a numerical analysis of Eqs.(1.1) for two-mode and three-mode operations. In both cases we take b_{ij}=b for simplisity.

4.1 Two-Mode Operation

First we follow the line $D_{10} = D_{20}$ in the parameter plane of Fig.2 to analyze the behavior of the solution of Eqs.(2.1). At the point $(D_{10}, D_{20}) = (1,1)$ a periodic solution emerges from the zero solution, it oscillates harmonically, with a small amplitude near the point (1,1). When D_{10} and D_{20} increase, the solution remains periodic and stable with an increasing oscillation period up to $D_{10}=D_{20}=19.09231$ at which the oscillation period tends to infinity. Beyond this point the only stable solutions are either $I_1 = 0, I_2 = I_{2+}$ or $I_1 = I_{1+}, I_2 = 0$, as is expected from section 2.

Based on the numerical integration of Eqs.(2.1) for numerous choices of the parameters, we have not found chaotic solutions, and the only time-dependent solutions observed are periodic functions for both two modes whose maxima appear alternatively. In particular, in the symmetric case (k_1=k_2 and D_{10}=D_{20}) both modes I_1 and I_2 have the same pulsation pattern. If f(t)=f(t+T) stands for a periodic pulsation of the mode 1, then the pulsation of the mode 2 is described by f(t+T/2) where T is the pulsation period.

4.2 Three-Mode Operation

We first study a situation corresponding to a symmetric choice of the parameters that are taken to be $k_1 = k_2 = k_3 = k = 5$, $D_{10} = D_{20} = D_{30} = D_0 = 40$ and $b_{ij} = b = 0.25$. The symmetric solution

$$I_1 = I_2 = I_3 = I = -A\pm(A^2+(D_0-k)/10g)^{1/2}$$
$$A = \frac{2bk+5g}{20g} \tag{4.1}$$

has been proved to be stable when $g < g_H$ ($=2.593$) where g_H is related to Hopf bifurcation I_H ($=0.9149$) through Eq.(3.4). Another steady-state solution is

$$I_i = I_{i+}, \quad I_j = I_k = 0 \tag{4.2}$$

with i,j and k different of one another. This solution is proved to be stable when $g > \tilde{g}$ ($=10.11$) where $\tilde{g}$ is the positive root of the following equation

$$D_0 = \left(1+b\frac{-k-g+\sqrt{(k-g)^2+4gD_0}}{2g}\right)\left(\sqrt{(k-g)^2+4gD_0} - g\right)$$

With these informations obtained analytically we study only the domain $g_H < g < \tilde{g}$ where no steady-state solutions are stable. Eqs.(1.1) with N=3 were integrated numerically. For arbitrary initial conditions, the final state reached was observed to be

$$I_i = f_1(t), \quad I_j = I_k = f_2(t) \tag{4.3}$$

with i,j and k different. Here $f_1(t)$ and $f_2(t)$ are two periodic functions that have the same period and are out of phase with each other, as shown in Fig.3. Near the point $\tilde{g}$, both $f_1(t)$ and $f_2(t)$ have a quasi-steady state (see Fig.3(E)), the duration time of quasi-steady state of f_1 is much greater than that of f_2. When g tends to $\tilde{g}$, they will be infinite and zero, respectively.

Another domain that we have studied is a non-symmetric choice of the parameters: $k_1 = 10$, $k_2 = 12$, $k_3 = 14$, $D_{10} = 150$, $D_{20} = 180$, $D_{30} = 140$ and

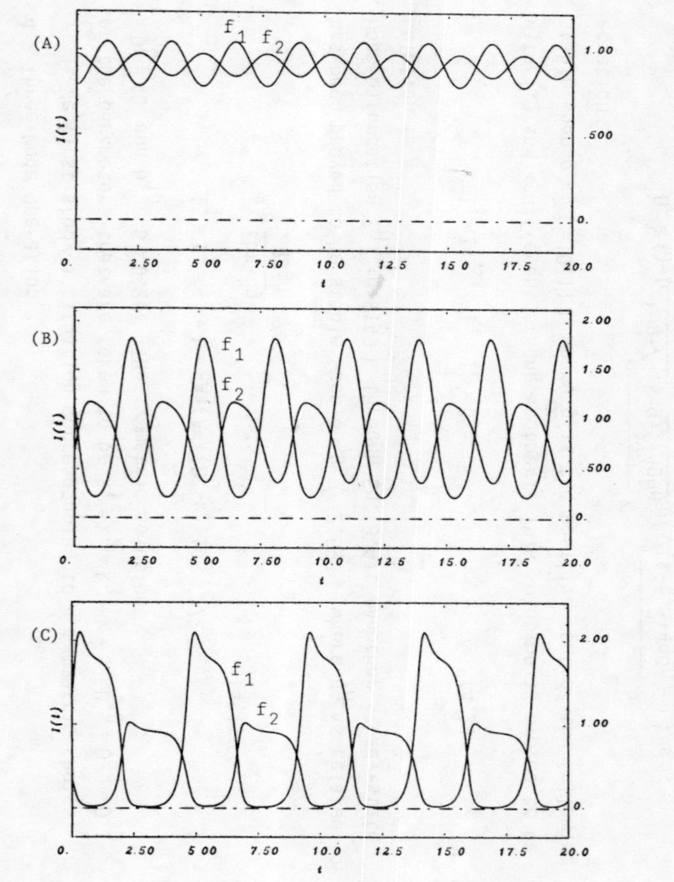

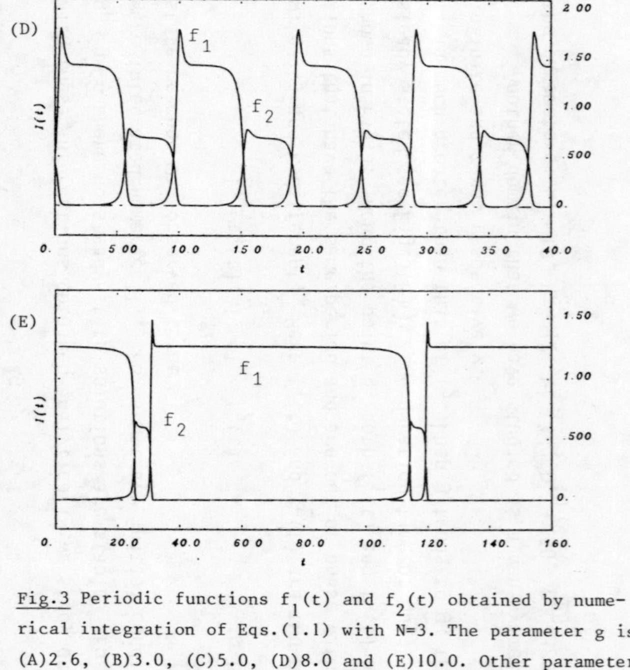

<u>Fig.3</u> Periodic functions $f_1(t)$ and $f_2(t)$ obtained by numerical integration of Eqs.(1.1) with N=3. The parameter g is (A)2.6, (B)3.0, (C)5.0, (D)8.0 and (E)10.0. Other parameters are $k_1=k_2=k_3=k=5.0$, $D_{10}=D_{20}=D_{30}=D_0=40$ and b=0.25. For arbitrary initial condition, the final state reached is always

$$I_i=f_1(t),\ I_j=I_k=f_2(t)$$

with i,j and k different of one another. When g tends to $\tilde{g}$ the period of $f_1(t)$ becomes infinite.

b=0.25. In this case the observed solutions exhibit a rich structure. These solutions are:

(1) For $0 < g < 1.7$, steady-state solution $I_1 > 0, I_2 > 0$ and $I_3 > 0$ i.e. all the three modes are above their threshold and their output is steady.

(2) For $1.7 < g < 3.8$ a hierarchy of period-doubled solutions was observed. The periodic solution emerges from the steady-state solution via a Hopf bifurcation at $g = 1.7$. Its period is first doubled at $g = 3.4$, then it is doubled again and again as g increases. At $g = 3.8$ the period tends to infinite.

(3) For $3.8 < g < 7.9$ the observed solution is chaotic. The maximun values of mode 3 becomes smaller and smaller as g increases.

(4) For $7.9 < g < 18.6$ the mode 3 in entirely inhibited and modes 1 and 2 are periodic.

(5) For $g > 18.6$ both modes 1 and 3 are inhibited and mode 2 takes the steady state I_{2+} defined by (2.4).

It should be noted that the steady-state solution $I_2 = I_{2+}$, $I_1 = I_2 = 0$ is not the only stable solution for $g > 18.6$. A linear stability analysis for the single-mode steady states

$$I_1 = I_{1+},\ I_2 = I_3 = 0 \tag{4.4}$$

$$I_2 = I_{2+},\ I_3 = I_1 = 0 \tag{4.5}$$

$$I_3 = I_{3+},\ I_1 = I_2 = 0 \tag{4.6}$$

is easily performed and leads to the following conclusions:

(i) For $18.62 < g < 69.53$ the only stable solution is (4.5).

(ii) For $69.53 < g < 113.11$ there are two stable solutions (4.4) and (4.5). Thus $69.53 < g < 113.11$ is a bistable domain.

(iii) For $g > 113.11$ all the three solutions (4.4), (4.5) and (4.6) are stable. This is a tristable domain.

ACKNOWLEDGEMENT

I would like to extend my thanks to Prof. Paul Mandel for his guidance. This work was supported by the Commissariat Général aux Relations

Internationales (Belgium) and the Solvay International Institutes of Physics and Chemistry.

REFERENCES

1. P.A. Franken, A.E. Hill, C.W. Peters and G. Weinreich, Phys. Rev. Lett. 7, 118(1961).
2. P.D. Drummond, K.J. McNeil and D.F. Walls, Optica Acta 27, 321(1980); 28, 211(1981).
3. P. Mandel and T. Erneux, Optica Acta 29, 7(1982).
4. X.-G. Wu and P. Mandel, J. Opt. Soc. Amer. B 2, 1678(1985).
5. P. Mandel and X.-G. Wu, J. Opt. Soc. Amer. B 3, 940(1986).
6. T. Baer, Laser Focus/Electro Optics 6, 82(1986).
7. T. Baer and M.S. Keirsead, paper ThZ1 at the CLEO 85 Conference.
8. T. Baer, J. Opt. Soc. Amer. B 3, 1175(1986).
9. X.-G. Wu and P. Mandel, " Second harmonic generation in a multi-mode laser cavity ", J. Opt. Soc. Amer. (to appear).
10. G. Ioos and D.D. Joseph, Elementry Stability and Bifurcation Theory (Springer-Verlag, New York 1980).

EFFECTS OF CURRENT DENSITY ON ETCH RATES OF SEMICONDUCTOR MATERIALS

Wei-Xi Chen
Physics Dept. of Peking Univ.
L.M.Walpita, C.C.Sun and W.S.C.Chang
Department of Electrical Engineering and Computer Sciences
University of California at San Diego

ABSTRACT

Ion beam etching are important processes for precision fabrication of device structure in micron and submicron range. They are applicable in optoelectronic device such as bistable optical devices.

We report reactive ion beam etching of semiconductors using gases, Such as CF4 and C2F6, and compare the etching rates with those of argon ion milling. For argon ion milling the etch rates increased linearly with current density. However for C2F6 and CF4 the etch rates increased with current densities at low current density but decreased with increase of current density at high current denisity; evidence suggests possible reason to be carbon accumulation on the semiconductor surfaces.

We have investigated the etching rates of InGaAs, InP, GaAs, Si, Ge as a function of ion beam current density using reactive gases,such as CF4, C2F6,and compared these etching rates with the etching rates of insulating materials.We show that nonreactive ion beam milling (Ar) of these semiconductor materials is much faster than reactive ion beam etching. However, reactive ion beam etching gives better ratios of photoresist/substrate etching rates than nonreactive etching. The etching rates using C2F6 and CF4 increase with increase of current density at low current density, but decrease with increase of current denisity at high current . There is some evidence to believe that this might be caused by carbon accumulation on the surface of the structure[1,2].

The ion beam etching process has been carried out in a commercial ion milling machine [3] (Millatron IV manufacture by the Commonwealth Scientific Company). A description of the system, sample preparation and etching procedure is given elsewhere[3].

1. Comparison of the etching rates for different materials

Figure 1a shows the etching rates of InGaAs, Inp, GaAs, Si and Ge as a function of ion milling current density for pure Ar. It is obvious that for these samples the etching rate is linearly proportional to ion beam density. The etch rates per unit current density of Inp, GaAs, Si, Ge is given in Table I.

Table I

sample	InP	GaAs	Si	Ge
Å/min ma cm	1300	1170	367	333

The samples have also been etched using CF4 and C2F6 gases. The current denisity dependence of these etching rates are shown in Figures 1b and 1c. The etching rates increase with an increase in current density in the low current density range (less than 0.4ma/cm for CF4 and less than 0.55 ma/cm for C2F6) but decrease rapidly with increase of current density in the higher current density range. This is in contrast to what is obtained in Ar ion milling.

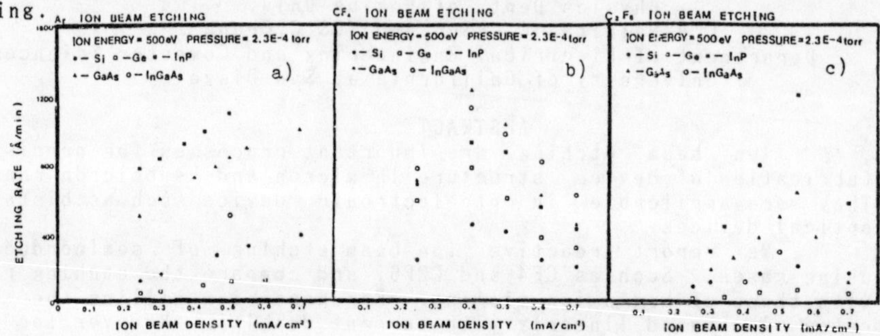

Fig. 1. The etching rates of semiconductors as a function of ion beam density for (a) argon (b) CF_4, (c) C_2F_6.

In order to obtain some insight to the effects of lowering the reactive etching rate of semiconductor materials at high current densities, we have measureed the etching rates of oxides such as SiO2. For SiO2 the dependence of etching rates on current density is shown in Figure 2. It is intersting to note that the etching rates are higher for C2F6, CF4 than for argon. The etching rate dependence on current density is also more linear for the oxides.

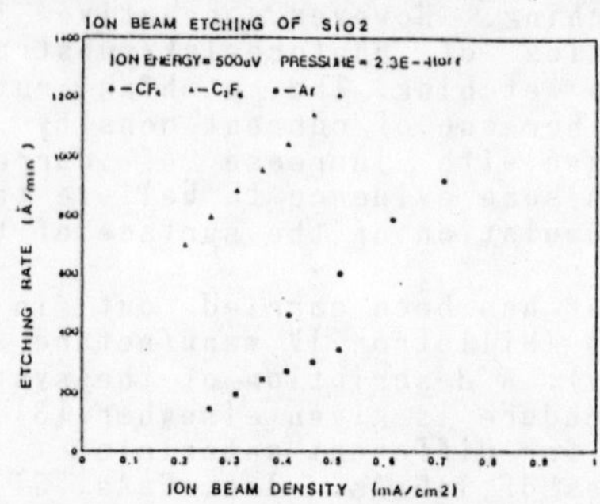

Fig. 2. The etching rates of SiO_2 for gases CF_4, C_2F_6 and Ar. The chamber pressure - 2.3 x 10^{-4} torr and ion energy -500 eV.

Obviously, a high differential etching rate between materials and photoresist is very useful for the fabrication of electronic and optical devices.We have seen that the differential etch rates were higher for C2F6 and CF4 than argon. For example, C2F6 etch rates gave 3 time for Inp, 2 time for GaAs, and 1.5 times for Si in comparison with argon ion milling.

2. Effect of Magnetic Field and Oxygen

It has been shown that the increase of the axial magnetic field causes percentage carbon in the beam to increase[1]. This implies that the percentage carbon on the substrate surface could also correspondingly increase with increase of magnetic field and alter the etching rates.Previous work on silicon has indicated that the lowering of etching rates of silicon at higher current

densities could be caused by carbon accumlation of the etched surfaces[1,2]. Therefore, in order to further verify the anomolous behaviour of the etching rate of semiconductors at high current densitiies,we studied the effect of increase of magnetic field in the plasma chamber on the etch rates but held the current density constant; the etching rate decreased with the increase of magnetic field for the semiconductors implying increased carbon on the surface.but it increaswd with increase of magnetic field for the photoresist resulting in a reduced differential etch rate [Table II].
In the case of oxides,as seen in Fig.2 etching rates are better for C2F6 and CF4 than Ar. Here the oxygen in the lattice appears to form volatile compounds such as CO, CO2 or COF2[4] and therefore the amount of carbon on the surface should be reduced. As an additional experiment we used a gas mixture consisting of 80% C2F6 and 20% of O2 to etch InP. For the same etching conditions and at high current densities the etch rates were higher than those given by 100% C2F6, indicating the possibility of reduced carbon in the presence of oxygen. Dependency of the etching rates of semiconductors on the current density can therefore be altered both by altering the magnetic field and by introduction of oxygen into the chamber.

Table II

Beam Current Density mA/cm	Magnet I (A)	Magnet V (V)	InP Etching Rate Å/min	AZ1350 Etching Rate Å/min
.5	1.5	1.7	1200	80
.5	2	2.5	440	100
.5	3.4	4.1	250	120

Nonreactive etching rates of most substrates showed linear dependency of the current density. Etching of oxides also showed linear dependency of current density for both nonreactive and reactive etching. It has been shown previously[1], that the lowering of etch rates in silicon at high current densities under reactive etching were caused by carbon accumulation. We have shown some evidence as to the effect of carbon on etching rates of semiconductors by adjusting the magnetic field in the plasma chamber and also by adjusting the oxygen percentage in the plasma.The reduction in magnetic fields caused the substrate etching rate to increase indicating that the carbon accumulation is reduced. Similarly,introduction of oxygen into the plasma chamber also caused the substrate etching rates to increase. The etching rates could therefore be controlled by adjusting both the magnetic field and the oxygen percentage.

REFERENCES

1. T.M.Mayer and R.A.Barker, J.Electrochem. Soc. 129,585(1982)
2. M.Miyamura, O. Tsukakoshi and S.Komiya, J.Vac.Sci.Technol.20,986(1982)
3. Wei Xi Chen, L. M. Walpita, C.C.Sun, W.S.C.Chang, J.Vac.Sci.Tech B4(3),May/Jun p701-705 1986
4. J.W.Corbun, H.F.Winters and T.J.Chuang, J.appl.Phys.48,3532(1977)

Summary of the Panel Discussion

K. K. Lee and H-Y Zhang

The following is a summary of the panel discussion held on Wednesday afternoon, August 26, 1987. The panel members were H. Gibbs, I. C. Khoo, and C. Seaton.

For continuity, we tried to group the subject matter discussed at different times together. The subjects were either brought out by the panel members or by the audience. The subjects are not listed in their chronological order. We are fully responsible for any inadvertent omission and inaccuracies of this summary.

The panel was assembled to facilitate and devices which might be used for nonlinear decision making and optical computing. Here optical computing is taken in a broader sense, i.e., it not only means numerical digital computing, but also includes other applications of non-linear devices for decision making.

Gibbs first asked the question: for parallel optical computing, what material one might intend to use for nonlinear decision making? It seems that GaAs/AlGaAs multiple quantum well (MQW) is the material favored by many. The reasons for choosing GaAs to build an optical decision making device are: its large nonlinearity, its ratio of power to recovery time is competitive to any other material known, it is compatible with diode lasers, it can be integrated with electronics,and it can operate at room temperature. Physical characteristics and mechanisms of GaAs as well as those of InSe were briefly reviewed by Gibbs. Physical characteristics of ZnSe,

ZnS, CdS, etc. were also briefly summarized. Gibbs also pointed out that comparing n_2 of a candidate material with n_2 of GaAs is not meaningful, because n_2 of GaAs is not an accurate description; the change in index does not follow the intensity like n_2I but also depends on the carrier density. That is why there is so much confusion about GaAs's nonlinearity and comparisons with other materials. Moreover, since the nonlinearity is a resonant interaction, it depends on wavelength. It also saturates, i.e., dn/dI decreases as I increases.

At this point Mandel pointed out that one could not discuss what the best material would be. Such question becomes meaningful only after one has decided what the application is.

Gibbs agreed and stated that he had already given the goal of parallel optical computing. He pointed out that when one talks about parallel optical computing, one needs to make decisions, to cascade, to operate at low power, etc. One might not even be able to do all these. Unfortunately, it is still a prevalent concept that optical computing device is limited to a numerical optical computer trying to compete with a digital electronic computer. Since electronic array processors already exist which perform 10 E14 bit operations per seconds, can optics play any role? The consensus of the panel was that 'optical computing' will not be strictly numerical digital computing operating like an electonic digital computer. Nonetheless, one may have some hope in the global interconnectivity. The latter maybe especially important for neural networks or associative memories, but most demonstrations are rather trivial at present.

Note that, in a neural network, speed is not the primary consideration. Low-power operation with a material such as a photorefractive material or liquid crystal for image processing could be quite important.

As discussed in the morning, the room-temperature band-edge-resonance-enhanced nonlinearities for semiconductors are due to exciton renormalization, band filling, resonance enhancement, etc; all can be comparable, but different mechanisms dominate different materials. Nonetheless, from a practical point of view, the nonlinearities are about the same magnitude.

On the wavelengths of the devices, unlike optical data storage or optical communication, where one finds preferred wavelengths for such applications, there is no preferred wavelength for optical bistability.

There were further discussions and speculations on how to increase the nonlinearity and lower the operating optical power. Quantum dots were suggested; their difficulties were pointed out, e.g., uniformity in size and sufficient density. Also, shrinking the device down is another way to reduce the energy required. Another way of reducing the energy required is to deal with slower devices but with massive parallelism. Indeed, most parallelism demonstrations are using liquid crystal light valves, with recovery time of 0.1 second.

Symbolic substitution was briefly explained and its difficulty in practice was also discussed.

From the nonlinear decision making point of view, another attractive approach is nonlinear serial processing of data, rather than switching of routing. One way is to use nonlinear waveguides. The discussion then turned over to Colin Seaton.

Seaton then discussed the material requirements for waveguide signal processing which are: high nonlinearity (>10E-7 esu), fast speed (<100 ps response time) with 10mW peak power. Whether these can be achieved by resonance enhancement depends on the damage threshold due to resonance enhanced absorption. Ideally, one would like to use nonresonant nonlinearity, if attainable. Fast response can be achieved (<1ps), but may need some kind of resonance effect to obtain large nonlinearity. For serial processing, both switch-on and -off have to be about 1ps in order to have large bit operation rate. Furthermore, the material must be able to withstand the damage threshold due partly to the short pulse and high peak power, and partly also due to the intrinsic requirement of the waveguide to be low-loss material (<1db/cm), One also needs proper format and processability of the material, for the desired applications. Thus material fabrication becomes a large part of the problem. In this context, GaAs-based materials have some added advantages due to semiconductor industry experience of this material.Such processing technology and experience will be necessary for integrated optics.

One often hears that some organic materials have large nonlinearity, some others have fast response, or some can meet other critical performance parameters.But it should be pointed out that satisfying one or two performance parameters is not sufficient, the required material has to satisfy all the parameters. Organic crystals and polymers are fairly stable. In order to have a large damage threshold, the material has to be in a very pure form. A few materials in the bulk have GW/cm^2 damage threshold. Nonetheless, damage threshold is a major problem for organic material. Suffice it to say that organic materials still need further study in order to make possible a proper assessment.

For waveguide applications, the relevant parameters are: nonlinearity, response time, absorption coefficient (including detuning from resonance), and saturation value of the index change one can achieve. It is clear that GaAs MQW scores quite well in all the categories. Of course, the choice of the material depends very much on the application.

As far as utilizing an optical fiber as a nonlinear waveguide is concerned, it has a small nonlinearity, but since it is a Kerr medium, it has a very fast response time. Also, it has a very low absorption coefficient, indeed nearly lossless. Thus it also has very high figure of merit. It is a very active area of investigation. One interesting approach is to change the chemical composition of the glass to increase the polarizability of the glass, i.e., increase the nonlinearity of the glass without reducing other virtues.

Another area of interest is semiconductor-doped glass. There, one wants to get material which has best of all the virtues. For instance, it has the well-established processability of glass technology, yet it has resonance enhanced nonlinearity of the doped semiconductors in the glass. Furthermore, semiconductor-doped glasses not only have fast switch-on, but also have fast switch-off due to very short surface recombination time.

And with those small (~10 Å) crystallines, the carriers can not diffuse, thus one gets around the usual long bulk diffusion time. Consequently, one can get devices with higher modulation rate due to the fast decay time.

Concerning materials, there are two interesting questions which need to be answered. First, what are the fundamental limits and how to achieve those limits in materials which are better than the existing ones. Second, from a practical point of view, with the existing materials, how can one best utilize them. Both approaches need to be explored.

Prof. Enderlein of Humboldt University (E.Berlin, G.D.R.) gave a brief review of both theoretical and experimental work on optical bistability at his institution. Questions on thw details of the experiments then followed. Questions then came back to ZnS, ZnSe, whether exciton had been observed for these material at room temperature. It was pointed out that for ZnSe it had been observed.

It was asked that if n x n array was not 100% reliable, what would one do. Gibbs' view is that a uniform 1 cm^2 2-D array would not solve any important technological problem. In fact, such a 2-D array will certainly come into being shortly. Uniformity is not a fundamental problem, but a growth process. Note that an n x n array still needs to be coupled to other devices, i.e., needs to be interconnected, etc. to make it a useful device.

Khoo described a fiber coupling scheme via self-focusing or self-defocusing of a nonlinear medium, which in turn changes the apparent of these fibers. It was claimed that as little as pJ energy will be sufficient for changing the coupling coefficient between fibers. Khoo then talked about liquid crystals as a nonlinear material. He pointed out that it responds very well in the IR region. Liquid crystals can also be used for IR image processing, IR light valve, modulator, IR self-pumped phase conjugator, etc.

Questions were then asked about optical interconnects in optical computing. Gibbs' comment was that optical interconnects will make a contribution in optical computing or optical decision making. He speculated that neural network type of work will center on optical interconnects with good electronics.

Finally, there were questions on the architecture and systems of a digital optical computer. To these questions, Gibbs' comment was 'No one knows how to build a digital optical computer which is competitive with a digital electronic computer'.

AUTHOR INDEX